WORKING WITH
MATERIALS
WOOD • METAL • PLASTIC

Colin Chapman
Regional Adviser,
The Engineering Council's Technology Enhancement Programme

Mel Peace
Head of Technology,
De Aston School, Market Rasen

with

Gerry Denston
Advisory Teacher,
Manufacturing and Technology

and

Val Charles
National Curriculum Assessment Co-ordinator

Collins Educational
An imprint of HarperCollinsPublishers

TECHNOLOGY ENHANCEMENT PROGRAMME

The Collins Real-World Technology *series is approved by the Technology Enhancement Programme.*

Full details of the TEP can be obtained from:

TEP
5 Old Mitre Court
London EC4A 1YY

Tel: 0171 583 0900
Fax: 0171 583 0909

MIDLAND EXAMINING GROUP

The Collins Real-World Technology *series is approved by the Midland Examining Group to support the teaching of their GCSE Design and Technology syllabuses.*

MIDLAND EXAMINING GROUP
SYNDICATE BUILDINGS
1 Hills Road
Cambridge CB1 2EU

Tel: 01223 553311
Fax: 01223 460278

The Authors

Colin Chapman has worked in industry and education as a trainer, teacher and adviser. He has previously co-authored *Collins CDT: Design and Realisation, Collins Technology for Key Stage 3* series and BBC Enterprises *Techno*. He is currently Chief Examiner for MEG GCSE D&T: Engineering, an Assistant Examiner for ULEAC A-level Design & Technology, and a member of the DATA Advisory group for Secondary D&T examinations.

Mel Peace has considerable experience in the teaching and development of design & technology. He is an A-level examiner for ULEAC, and a member of the SCAA development group for the A/AS level core in design & technology. He has previously co-authored *Collins CDT: Design and Realisation*.

Published in 1996 by Collins Educational
Reprinted 1996, 1997 (twice), 1998
An imprint of HarperCollins*Publishers*

77-85 Fulham Palace Road
Hammersmith
London
W6 8JB

ISBN 0 00 327351 2

Designed by Ken Vail Graphic Design (production management Chris Williams)
Cover Design by Ken Vail Graphic Design
Cover photographs: The Image Bank, Peter Sharp and Telegraph Colour Library

Illustrated by Karen Donnelly, Simon Girling & Associates (Graham Bingham, Alex Pang, Stephen Sweet), Graham-Cameron Illustration (Jeremy Bays, Tony Dover), Ken Vail Graphic Design

Location and studio photography by Peter Sharp

Printed and bound by Printing Express Limited., Hong Kong.

Series planned by: Graham Bradbury and Alison Walters
Commissioning Editor: Alison Walters
Copyeditor: Janet Swarbrick
Production: Sue Cashin

Contents

1·Manufacturing perspectives

The manufacturing of products today involves new technology and design ideas in order to implement, design and make products that are marketable and can be sold to create wealth. A manufacturing base is crucial to the prosperity of most countries. Although some countries have an abundance of easily accessible natural resources, such as oil and minerals with which they can trade, many developed countries rely upon the sale of manufactured products for international trade.

Fig. 1.1 *The Industrial Revolution created hardship for many, but made Great Britain the world's wealthiest nation.*

The manufacturing base in Great Britain developed in the eighteenth century, during the period known as the '**Industrial Revolution**'. Many important developments at this time contributed to Great Britain becoming the world's most prosperous nation. For instance, the development of factories, particularly within the textile industry, replaced small 'cottage' industries, while the invention of the steam engine enabled the factories to utilise cheap power wherever they were located. A plentiful supply of iron ore and coal meant that it was possible to produce large industrial machines to improve manufacturing efficiency. Another very important factor was the development of an infrastructure of roads, railways and canals so that raw materials and manufactured products could be easily transported to their destination.

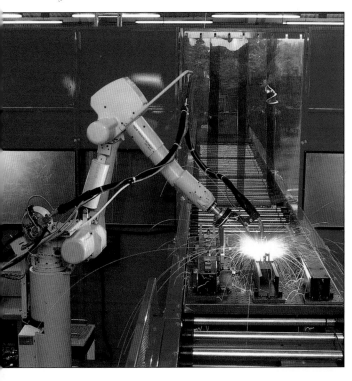

Fig. 1.2 *Modern manufacturing using the latest technology*

A wealthy nation is able to support a high standard of living, for instance by building new schools and hospitals. It is able to build and maintain the roads, railways and communication networks that in turn contribute to the success of the manufacturing industry. This chapter looks at how the modern manufacturing industry is organised and how it makes use of new technologies in order to continue to be competitive in the world market and succeed in creating wealth.

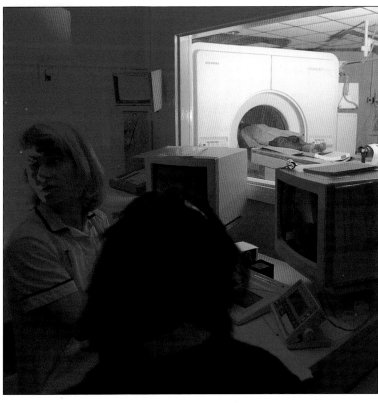

Fig. 1.3 *Wealth can be used to raise a country's standard of living and provide better services*

MANUFACTURING AND THE ECONOMY

'Our aim is commercial and industrial success and the rewards it will bring for all of us. To achieve this we seek a new partnership between government and industry.' (John Major, The Prime Minister, 1994.)

Fig. 1.4 *The logos of some of the world-class UK manufacturing companies*

The **Confederation of British Industry (CBI)**, in a survey of the best practices used in British manufacturing companies, found all aspects of world-class manufacturing, including quality, cost, delivery, innovation, marketing, design and environmental awareness. These were combined with a sustained drive for continuous improvement. Most important however, was the recognition that *people* remain central to the achievement of goals. Teamwork and flexibility within multi-disciplinary teams is a key factor in this. The ability of manufacturing companies to bring about change with the full backing of the workforce sets them apart from those that just have good intentions. Forty-five of the world's top manufacturing companies are UK owned, which is more than any other European country.

Manufacturing in the UK has also been bolstered by the investment of Japanese companies such as Nissan, Honda, Mitsubishi, Epson and Toyota. They have brought jobs to areas where unemployment has been high and created a demand for components from local manufacturers. They have stimulated local economies and encouraged the growth of service industries. As important as their influence on British manufacturers is, productivity in the majority of

Fig. 1.5 *The Nissan factory in Sunderland*

British manufacturing companies still lags behind the best of our continental competitors. We will see later how this is being tackled by investment in information technology and the adoption of the best production techniques from Japan and the USA.

The CBI identifies twelve manufacturing sectors in its report *Making it in Britain 3*. The twelve sectors are represented in Figure 1.6, in the form of a pie chart, with their output as a percentage of total manufacturing output. The figures are for 1993. It is worth noting that almost 50% of manufacturing is concerned with the three basic materials, wood, metal and plastics. The UK creates 4% of the world's gross domestic product and accounts for 6% of world exports.

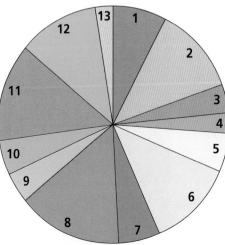

Fig. 1.6 *The relative sizes of the 12 manufacturing sectors*

1 Metals and metal products 7.7%
2 Chemicals 11.8%
3 Mineral products 3.9%
4 Wood products 2.9%
5 Rubber and plastics 5.2%
6 Paper, printing and publishing 12.0%
7 Textiles and clothing 5.7%
8 Food, drink and tobacco 14.9%
9 Aerospace 4.1%
10 Motor vehicles 4.8%
11 Electrical and electronic engineering 13.3%
12 Mechanical engineering 11.2%
13 Other 2.5%

Every year manufacturing earns approximately 65% of the UK's foreign earnings and it is estimated that each worker employed in manufacturing provides a job for at least one other worker in the service sector. The manufacturing industry plays a vital role in providing jobs and will continue to do so. In 1992–93 the total number of people in employment was 24,819,000. The bar chart (Figure 1.7) shows how this figure was spread across each sector of the economy.

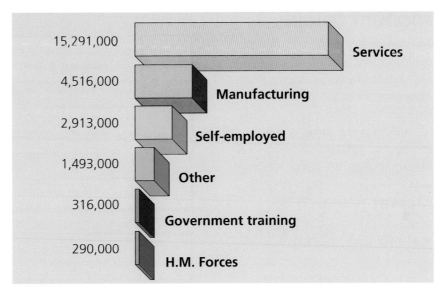

15,291,000	Services
4,516,000	Manufacturing
2,913,000	Self-employed
1,493,000	Other
316,000	Government training
290,000	H.M. Forces

Fig. 1.7 *How the workforce is divided up across the economy (1993)*

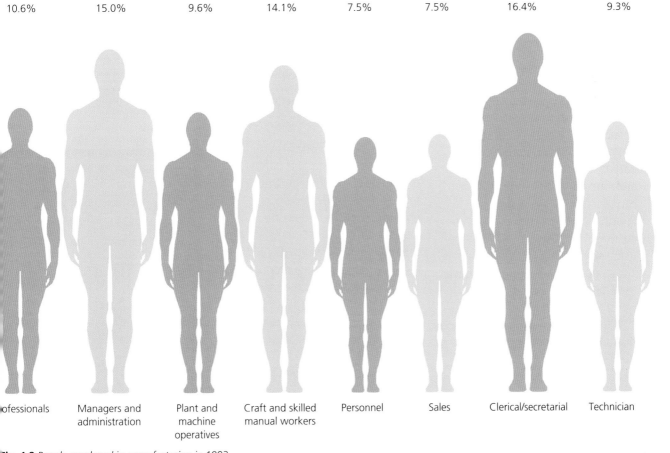

10.6%	15.0%	9.6%	14.1%	7.5%	7.5%	16.4%	9.3%
Professionals	Managers and administration	Plant and machine operatives	Craft and skilled manual workers	Personnel	Sales	Clerical/secretarial	Technician

Fig. 1.8 *People employed in manufacturing in 1993*

In some areas of concentrated manufacturing industry up to half the working population can be directly employed in manufacturing. In Walsall in the West Midlands, for example, it is estimated that a third of all those in employment will be in manufacturing for many years to come. Where a manufacturing industry has relocated to rural areas it can become the only source of employment. Manufacturing remains vital to both the national and local economies. The types of employment opportunities in manufacturing, however, continue to change. The table in Figure 1.8 shows the proportion of people employed in the occupation of manufacturing in 1993. It is important to note that the number of unskilled and semi-skilled people employed in the manufacturing industry has fallen in recent years, while at the same time the demand for professional, technical and skilled employees has grown. This is a trend that is expected to continue, with manufacturing industry demanding more and more technically qualified versatile people and needing fewer unskilled manual workers.

PRODUCT ORIGINS

The motivation for the manufacturing industry to develop new and improved products comes from a variety of sources. The requirements of the customer and the need to keep ahead of the competition exert a significant demand pull on the manufacturer.

Market research

Market research plays an important role in the development of new products. Manufacturers can draw upon useful information already available within their organisation. Sales and accounts departments will maintain records that can indicate which products and accounts are most profitable and will show if there is a pattern in the sales of particular products. Published information, particularly trade journals and government statistics, can show whether the demand for a type of product is growing, static or in decline, how the market is changing and who the competition is. After consideration of all the available information, it may be necessary to conduct market research out in 'the field', going directly to the consumer. The stages of market research that a manufacturing company might typically follow are shown in Figure 1.9. The benefit of market research is that it removes a lot of the guesswork from decision making thus reducing the element of business risk.

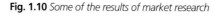

Fig. 1.9 *Stages of market research*

The three products shown in Figure 1.10 had interesting and different market research findings. Detailed market research by Kenwood enabled them to produce the Kenwood Chef. The Sinclair C5, on the other hand, was an idea of the inventor that became a product that nobody wanted. And the market research for the Walkman suggested that nobody would ever want a personal stereo system to carry around with them, which proves that experts do not always get it right.

Fig. 1.10 *Some of the results of market research*

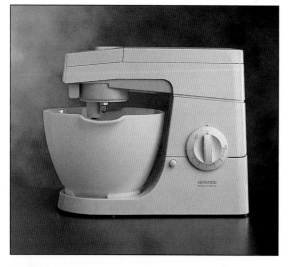

Technology push

Often, a new technology, material or process is developed for use in one particular context. This sets in motion a 'push' to apply it to other new and existing products, and is sometimes referred to as a solution looking for a problem! Many technological developments associated with aviation and space exploration, for example, have subsequently been applied to problems that existed in other fields of manufacturing, such as Teflon, a product now widely used to provide a non-stick coating to cooking utensils.

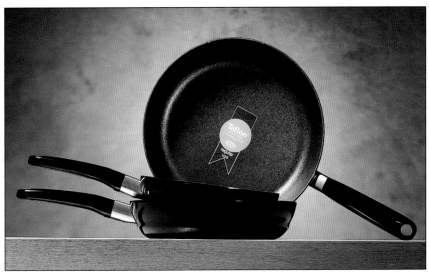

Fig. 1.11 *The Teflon used to coat these pans was originally developed for use in space!*

Polymers are continually being developed as cheaper, more efficient substitutes for metals. Polytetrafluorethene, for example, is an ideal new material for hip replacements. Newly developed plastic resins have revolutionised the manufacture of adhesives, which have subsequently been applied to the development of plywood and other laminates. The development of the means to mass-produce laminated chipboard and MDF (medium density fibreboard) has transformed furniture production and heralded the age of the 'flat pack'. As long as the demand from customers exists, new and improved products will continue to flow from the manufacturing industry which, out of necessity, will be looking for cheaper and more effective materials, processes and technology.

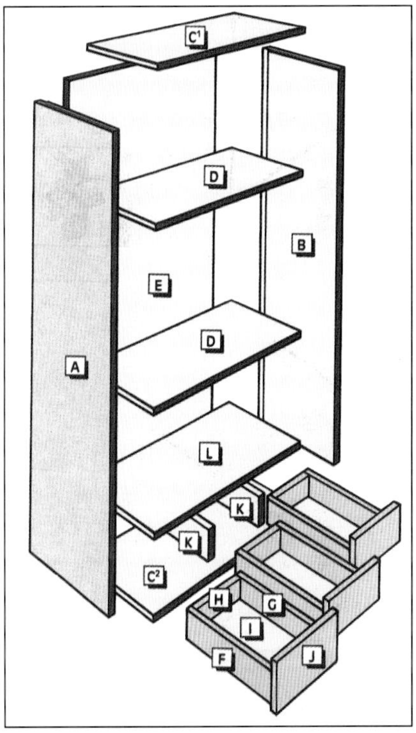

Fig. 1.12 *Flat-pack furniture*

Steps in the process of value analysis:

1 **Selection:**
   identify a product that is expensive for the function it performs, and that is made in high volume (which means it could therefore have good potential savings).

2 **Determination of function:**
   decide what the needs are of the product. What must it do to work effectively? When, where, how, why and by whom is it used?

3 **Collection of data:**
   include cost information, product drawings, specifications, data sheets, materials specification etc.

4 **Speculation phase:**
   analyse all the information in detail, considering all the alternatives.

5 **Evaluation:**
   test and evaluate proposals.

6 **Implementation:**
   manufacture the new and improved product.

Value analysis

During the Second World War the need to produce superior armaments from a limited supply of resources was viewed, literally, as a matter of life and death. A technique known as 'value analysis' was developed and applied to products by teams of designers and engineers and is still employed today. Value analysis is the systematic investigation of a product and its manufacture to reduce cost and improve value. Value is expressed as '**cost value**' and '**use value**'. The six steps in the process, shown on the left, are very similar to the model of a design process.

DESIGN, DEVELOPMENT AND PRESENTATION

'A picture is worth a thousand words' is an old saying and a true one. The human eye can feed the brain with more information faster in the form of a picture than by any other means. The design and development of new products and the instructions for manufacture therefore make full use of graphical means of presentation.

Recognising the need for a new product can happen in various ways– from market research, as a response to competitors, through the necessity to update a product or by the identification of a defect in a current design. The design team must clarify what is required, before producing a **design brief** and a detailed **specification** that will identify important design features such as function, cost, quality, performance, aesthetics, size and life expectancy. When the task is clearly defined the design team can then consider every aspect and detail of the product. Each component part will be conceptualised, analysed, improved, re-designed, analysed again, and so on until a final solution is arrived at. In this way human ingenuity finds expression in innovative designs.

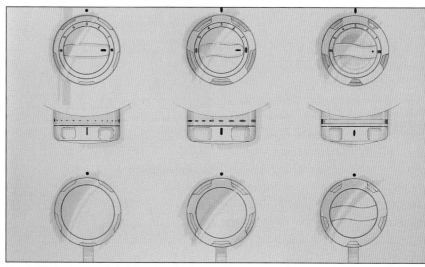

Fig. 1.13 *Development sketches for a bathroom shower control*

Computer technology can assist with all the design functions. At the Massachusetts Institute of Technology in 1963 Ivan Sutherland first demonstrated a **Computer Aided Draughting (CAD)** system; he called it 'Sketchpad'. The rapid development of the microprocessor has made it possible to create, modify and manipulate complex graphical images on a screen relatively quickly and store the data for future reference. Sutherland used a light pen as a pointing device. Modern CAD systems can use a variety of input and output devices, which are often referred to as 'peripherals'.

CADD (Computer Aided Design and Draughting) systems help the designer to develop ideas by enabling the creation of 2D (two dimensional) and 3D (three dimensional) solid and wire frame images on the screen. The images that are generated can be moved around and copied, and their shape and form manipulated and modified. The use of colour and animation enables the designer to display component parts more clearly and visualise how moving parts perform in relation to others.

Fig. 1.14 *Using a modern CAD system with a range of peripherals*

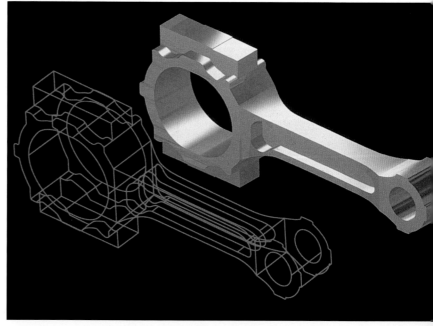

Fig. 1.15 *Wire frame and solid model computer generated images of a connecting rod*

Designers can be further assisted by software that performs a detailed engineering analysis of a product and its component parts. The effects of heavy loads or the reaction of a cutting tool to high temperatures, for example, can be simulated and analysed on screen.

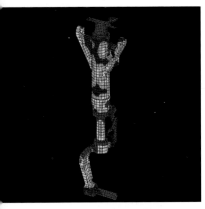

Fig. 1.16 *Computers can be used to simulate and analyse the effects of stress upon components.*

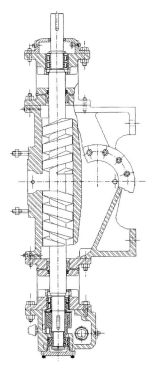

Fig. 1.17 *Computer generated sectional view of a rotary screw pump*

The final phase in this process is often the presentation of the design by means of working drawings that provide precise information about the product and how it is to be made. Computer generated drawings of this type will conform to the appropriate standards such as BS 308 and will have a parts list with quantities, material specifications, sizes, any special manufacturing requirements for each component part and any other information needed to manufacture the product.

One of the ways that CADD systems make these drawings easier and faster to produce is by accessing libraries of components and symbols that can be simply pulled into the drawing as required. Libraries can be modified and added to as more 'standard' elements are used.

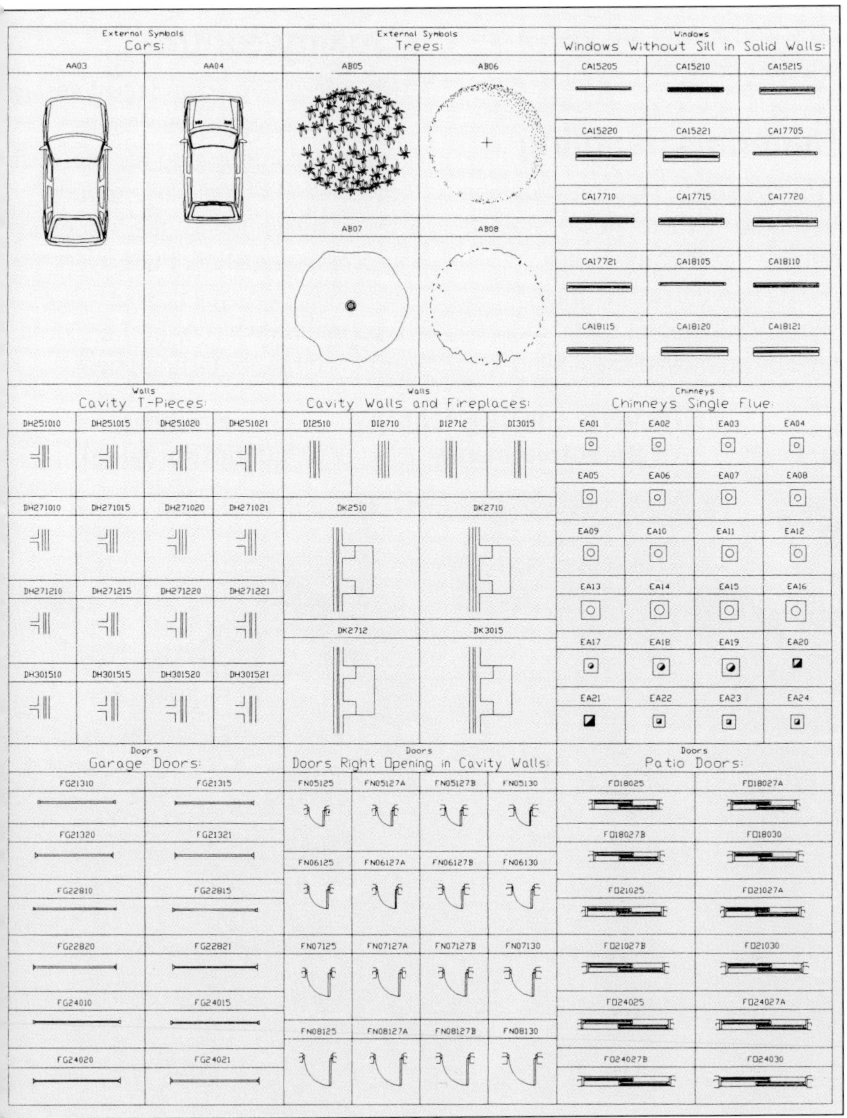

Advantages of CADD

- Drawings are generated up to three times faster than by conventional methods.
- The 'turn round' of design proposals and communication with the customer is quicker.
- The quality and accuracy of drawings is increased: plotters and printers produce superior text and line work, such as hatching and shading.
- The quality of design options is increased.
- The ability to conduct detailed analysis quickly reduces product development time and increases the variety of design alternatives.
- Access is gained to special draughting techniques such as rubber banding, rotation and translation, which are not available using conventional methods.
- It enables the storage and retrieval of precise design information quickly and easily.
- It allows integration with all the other contributing functions – manufacturing (see CADCAM, p. 17) marketing, sales, production planning, etc. (see CIM, p. 13).

Fig. 1.18 *Library symbols for architectural drawing using CADD*

PROCESS OPERATIONS

Manufacturing from raw material to finished product involves four basic functions: processing of materials; assembly of parts; materials handling; and quality control. Processing and assembly 'add value' to the product, and quality control ensures that quality is maintained and the product meets customer requirements.

Processing of materials

There are four main types of process operations (see Figure 1.19). **Primary** processes convert raw materials into the initial product form such as a cast aluminium engine block or an injection moulded compact disc case. **Secondary** processes give the product its final shape and form and may involve machining operations like drilling, turning and milling. Some operations do not change the physical appearance of a product but may alter the physical properties, such as heat treatment of metal parts to enhance hardness or toughness. **Finishing** processes are for protection or to enhance appearance such as painting, chromium plating, anodising.

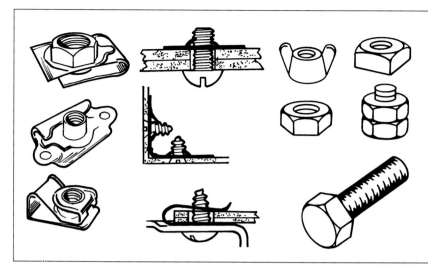

Fig. 1.20 *A range of mechanical fasteners used in assembly processes*

Assembly

The assembly of products involves two or more separate components being joined together. This will include mechanical fastening operations using screws, rivets, nuts and bolts, C washers and spring clips, and joining processes such as welding, soldering and gluing.

Materials handling

Without exception some provision has to be made to move materials between each stage in the production process. In many situations materials spend more time being moved than being

processed, and materials handling has therefore become a prime area for automation. Storage of materials between processes is avoided whenever possible.

Quality control

Manufacturing control must ensure that all the factory operations properly contribute to the successfu completion of a quality product. The management of the production process is dependent upon a constant, two-way flow of information which is best achieved using computer technology.

Primary (Casting)	Secondary (Machining)	Property enhancement (Heat treatment)	Finishing (Paint spraying)

Fig. **1.19** *The four main types of process operations*

COMPUTERS IN MANUFACTURING

The greatest contribution made by computers and computer technology to society is to help us to use our human resources more efficiently and, within the manufacturing industry, more productively. Productivity is the relationship between the quantity of components and products produced and the cost of producing them – the lower the cost the higher the productivity. Computerised manufacturing processes have contributed to higher productivity by reducing production costs. Initial investment costs are high and automation and computer control has sometimes meant that jobs are lost, but there can be benefits for those still employed. In many manufacturing organisations the three Ds of the Industrial Revolution – 'danger, dirt and drudgery' – have almost entirely disappeared and manufacturing companies are able to be more competitive, taking their place in world markets. The modern manufacturing unit is quite different from the factories of the past.

Fig. 1.21 *The 'danger, dirt and drudgery' of a nineteenth-century textile mill*

Computer Integrated Manufacturing (CIM)

Computer Integrated Manufacturing (CIM) uses computer technology to join together the various computer-assisted functions of a manufacturing company. Making what appears to be a simple product, such as a screw or a ballpoint pen, requires the work of a number of people, the majority of whom are not directly involved with the actual making. The diagram in Figure 1.22 shows the various functions that contribute to the manufacture of products ranging from a simple screw to a complex motor car. All the functions are coordinated by one central computer system. The computers are linked together like those on a school network. Each of the computer functions are performed separately but because they are linked to a central computer they are able to interact so that nothing is done in isolation. It is the software on the central computer that enables this interaction to take place. There can be a different type of software for each function. For example, if a design is modified using the CAD program then changes in the CNC (Computer Numerical Control) program that controls the machines will automatically take place. If a new order is generated then production planning and stock inventory are changed accordingly. This type of software is similar to popular integrated software packages like Microsoft Works, which can be found on many school networks.

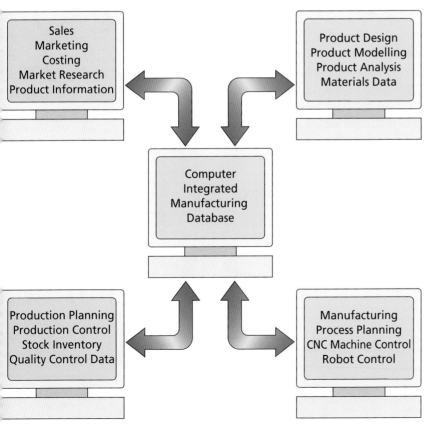

- Sales
  Marketing
  Costing
  Market Research
  Product Information

- Product Design
  Product Modelling
  Product Analysis
  Materials Data

- Computer Integrated Manufacturing Database

- Production Planning
  Production Control
  Stock Inventory
  Quality Control Data

- Manufacturing
  Process Planning
  CNC Machine Control
  Robot Control

Fig. 1.22 *Computer Integrated Manufacturing (CIM)*

Computer Numerical Control (CNC)

CNC machine tools are programmable, automated means of machining components. They are a development of the **numerical control (NC)** technique that first appeared in the 1950s which controls the actions of machines by the input of instructions in the form of a code. The coded instructions are supplied to the machine in blocks, each of which will result in a single operation. NC machines have 'hard-wired' electronic controllers which can read programmed instructions one at a time. A single line of instruction would contain several blocks like these: N010 G00 X350 Y250 M03 (the meaning of these codes is explained in Figures 1.27–1.29). A program containing the instructions to completely machine a component or part is made up of a large number of such blocks and is called a 'part program'. Early NC machines had no in-built intelligence so the part programs had to be loaded in manually, usually in the form of paper tape.

The paper tape was produced after a lengthy process undertaken away from the machine (see Figure 1.24). Working from a traditional paper drawing of a component, the part-programmer would work out the sequence of machining operations required and then write out each block of instructions by hand. The program would be transferred by keyboard to a paper-punch machine which converted each character into a row of punched holes in the paper strip. The completed tape was then taken to the machine and fed into its control unit. It had no memory and so could only read and act upon a single block at a time. After use, the paper tape was removed from the machine and stored until required again. The code used was **ASCII (American Standard Code for Information Interchange)** which uses seven-bit binary numbers to represent the alpha-numeric characters. Thus, on paper tape, a punched hole represents binary digit 1 and the absence of a hole represents binary digit 0. The block N010 would look like this:

```
N    1001110
0    0110000
1    0110001
0    0110000
```

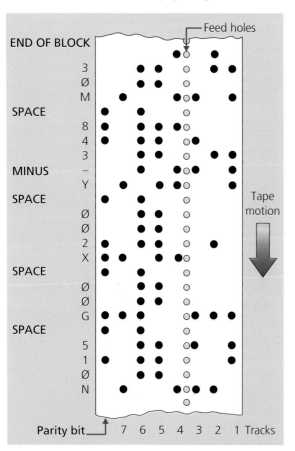

Fig. 1.23 *A piece of NC paper tape*

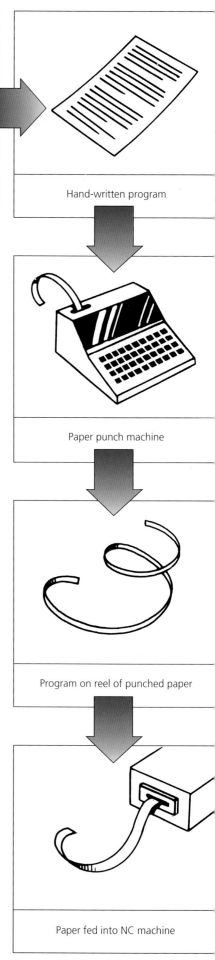

Engineering drawing

Hand-written program

Paper punch machine

Program on reel of punched paper

Paper fed into NC machine

Fig. 1.24 *Sequence for the production of paper tape*

CNC machines were properly developed in the 1970s. They maintained the basic principles of numerical control but now had a dedicated computer processor within the control unit. This enabled part programs to be stored in the machine's memory. Part programs can be installed in a variety of ways (see Figure 1.25): directly using a keyboard (keypad) on the machine, by paper, magnetic tape or floppy disk, or by data transfer from a remote computer which may be connected to several such CNC machines. This latter method is referred to as either **Direct or Distributed Numerical Control** (DNC) (see Figure 1.26).

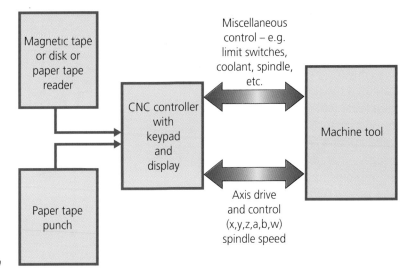

Fig. 1.25 *CNC machine block diagram*

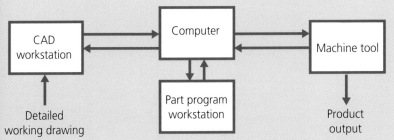

Note: the computer can directly control one or more machine tools at the same time

Fig. 1.26 *A direct numerical control arrangement*

Once the part program is stored in the machine's memory any number of repeat programs can be executed. A visual display unit (VDU) enables the operator to follow the part program through its cycles; some may even provide a graphical image of the tool path as the program progresses. CNC also allows direct communication with other computer systems such as **CAD databases** and **Computer Aided Production Management (CAPM)**.

Part programming

We have seen from the example N010 G00 X350 Y250 M03 that the coded instructions for CNC machines are made up of lines containing blocks. Each block is made up of a capital letter, called an address, followed by a number which together make up a complete command word.

Figure 1.27 shows the definitions of the letter codes.

G and M codes

The 'G' codes (see Figure 1.28) define the type of motion required. On a CNC lathe this would refer to the path of the cutting tool as it moves around the work piece. On a milling machine it applies to the work piece as it moves around the cutting tool although the cutting tool is able to move in both a vertical and angular direction on some machines. 'M' codes (See Figure 1.29) are miscellaneous functions such as starting the spindle and switching on the coolant.

Letter Code	Meaning
N	Line number in program
G	Preparation code – type of motion
M	Miscellaneous function – coolant, spindle off/on, etc.
X,Y,Z,A,B,W,I,J,K	Axis position in chosen units e.g. mm
F	Feed rate in chosen units e.g. mm/min
S	Spindle speed RPM
T	Tool number and length offset

Fig. 1.27 *Definitions of letter codes*

G00	Rapid tool movement	G33	Screw cutting
G01	Linear interpolation	G70	Inch units
G02	CW circular interpolation	G71	Metric units
G03	ACW circular interpolation	G80	Cancel canned cycle
G04	Dwell	G81–G89	Canned cycle
G17	Circular interpolation (X–Y Plane)	G90	Absolute coordinates
G18	Circular interpolation (Z–X Plane)	G91	Incremental coordinates
G19	Circular interpolation (Y–Z Plane)		

Fig. 1.28 *Common G codes*

M00	Program stop	M06	Tool change
M01	Optional stop	M07	Mist coolant on
M02	End of program	M08	Flood coolant on
M03	Spindle on (cw)	M09	Coolant off
M04	Spindle on (acw)	M10	Clamp on
M05	Spindle off	M30	Rewind tape

Fig. 1.29 *Common M codes*

Axes of movement

The direction in which slide movement is to occur in the tool in the lathe, or the work piece in the case of a milling machine, is defined by a letter (see Figure 1.30). This is most commonly X, Y and Z together with a (-) or a (+). (The (+) sign is not usually entered). Codes A, B and C are reserved for rotary movement around the three principle axes.

Fig. 1.30 *Direction of the slide movement on common machine tools – **a)** centre lathe, **b)** horizontal milling machine, **c)** vertical milling machine*

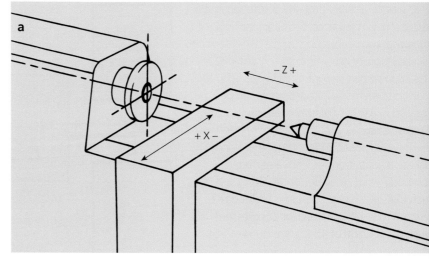

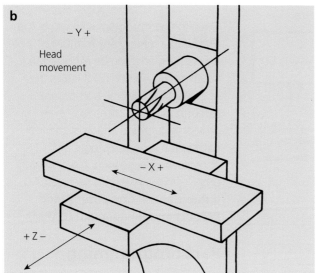

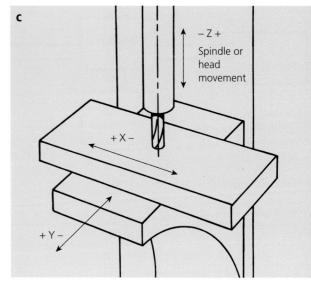

As shown in Figure 1.31, CNC lathes provide programmable tool movement in two axes, X and Z: this is referred to as 2D axis control. CNC milling machines can operate with either $2\frac{1}{2}$D or 3D axes control. The former is used for cutting simple slots or holes. 3D axes control provides programmable tool movement along three axes simultaneously and is used to machine more complex profiles. 5D axes control is also possible, the additional movement being provided by angular movement of the tool spindle axis. Complex sculptured machining operations with improved surface finish can be achieved on CNC machines with this capability.

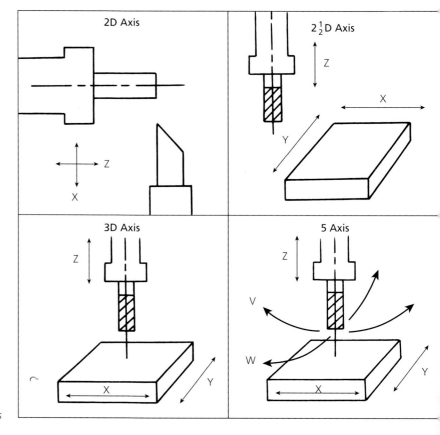

Fig. 1.31 *Control axes*

Absolute and incremental coordinates

Part programs contain all the instructions required to machine a component. The movement of the tool, or work piece relative to the tool, is expressed as linear coordinate values (XY) (XYZ). Consider the following line: N010 G00 X20 Y15 Z50 M03. Here G00 defines rapid tool movement to point 1 (X20 mm, Y15 mm); Z50 is the position of the tool 50 mm above the height of the work piece; M03 turns the cutting tool spindle on in a clockwise direction. The coordinates are distances measured from a pre-determined zero datum point (X0, Y0, Z0). This type of system uses absolute coordinates. An alternative system uses incremental coordinates where each point is specified by the distance moved from the last coordinate position. Modern CNC controllers allow both systems to be used within the same program simply by using the appropriate G codes – G90 or G91.

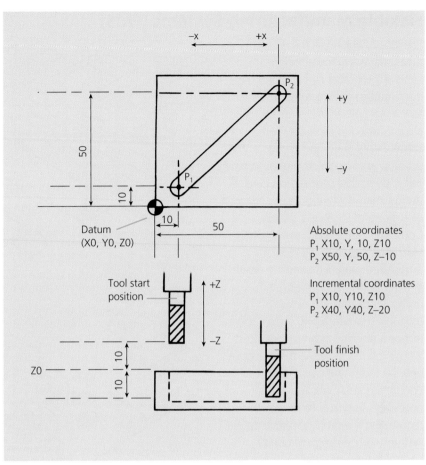

Fig. 1.32 *Absolute and incremental coordinates and zero datum point*

CAD/CAM

Computer Aided Design and Manufacture are two elements of the design and manufacturing process that are concerned with engineering functions. CAD systems provide the facility for product design, engineering analysis and detailed draughting. CAM includes process planning and CNC part programming. More recent software development has extended CAM facilities to include Material Requirement Planning (MRP), production scheduling, process control and production monitoring. CADCAM is a direct link between the design and the manufacturing functions. The ideal CADCAM system takes the design sps it automatically into a process plan for making the product. This includes the computer generation of part programs (Computer Aided Part Programming) and their automatic transmission to the CNC machine tools. The integration of all the engineering, business, planning and production control functions is discussed on page 13 under the heading 'Computer Integrated Manufacturing'.

Fig. 1.33 *CAD/CAM systems in use*

Flexible Manufacturing Systems (FMS)

The idea of linking all the different functions of a manufacturing company together by computer has been applied with increasing success to the workplace where the materials are processed to produce components, parts and products. The processing of materials is often carried out in dedicated **production cells**. A flexible manufacturing system consists of a group of such cells, usually Computer Numerical Control (CNC) machines, joined together by an automated materials handling system and controlled by an integrated computer system. Within a flexible manufacturing system there will be a team of technical operators whose job it is to set up the machines and the tools, maintain the equipment and make repairs as necessary. They are responsible for monitoring the process, the computer systems and carrying out quality control functions.

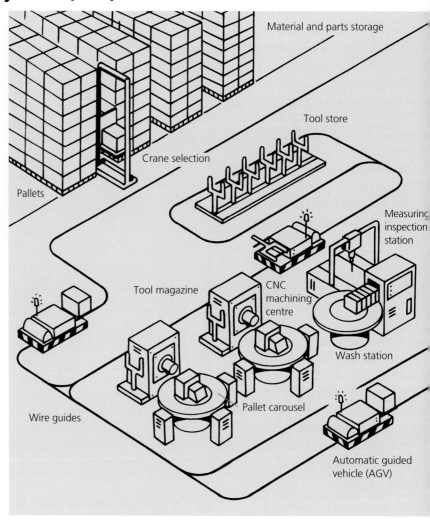

Fig. 1.34 *An FMS system*

Automatically Guided Vehicles (AGVs)

In many manufacturing organisations the movement of materials, components and tools between manufacturing centres and FMS cells requires a more flexible means of transportation than conveyors, tracks and overhead cranes. AGVs are carriages which can be programmed to travel along predetermined paths to specified locations. They are usually guided by inductive wires running on top of or below the factory floor and so are free to go wherever the wiring has been laid. Their routes can be programmed into an on-board microprocessor, known as a **programmable logic controller (PLC)**, which is able to respond to sensors incorporated in the vehicle. These enable it to communicate information regarding location, loading and potential collisions. AGVs may also be equipped with mechanical devices such as grab arms and telescopic forks.

Fig. 1.35 *AGVs in operation*

The types of machines used in FMS workstations will depend upon the products they are designed to manufacture. A visit to a machine tool exhibition and seeing the vast range of different machines now available is an exciting experience. Some manufacturers also employ engineers to design and build their own special purpose machines to meet the unique design features of their products.

Fig. 1.36 *A machine tool exhibition*

All the traditional material processing operations can be found in modern flexible manufacturing systems. A **CNC milling centre**, for example, is a single machine that can be programmed to face and end mill, drill, bore, counter bore and tap a thread. The tools required are mounted on a carousel alongside the machine and each tool is automatically selected in turn; the component is positioned and held as required for each operation. A **CNC turning centre** (Figures 1.37 and 1.38) combines all the operations of a traditional lathe in a single machine. The tools are mounted on a turret and the computer program instructs the machine to select the appropriate tool, sets the feeds and speeds, turns on the coolant, positions and holds the component, performs the operations and ejects the finished component. The turning centre in the photographs has two spindles and two 12-station tool turrets. After one turning operation the material is fed to the other spindle. It is separated with a parting tool and work commences on the other end. Both turning operations can take place at the same time.

Fig. 1.37 *A CNC turning centre*

The same CNC principles are now being successfully applied to material removal from sheet stock (i.e. pressing, punching, cropping, bending and cutting) as well as other process operations such as welding, forging, assembling, inspection and packaging. The results are greater accuracy, more consistent quality and greatly reduced waste. These and other processes may form part of a flexible manufacturing system.

Fig. 1.38 *Detail of tool turrets and drive spindles*

Robotics

Materials handling is a major consideration within all manufacturing processes and has an even higher priority within a system like FMS that represents a large investment of capital. Components and parts need to be transported to and between processes and on to and off machines. To maintain the productivity gains through process automation and integration, automated materials handling systems are being constantly developed. This is being achieved more and more by employing robot technology. The term robot comes from the Czech word 'robotnic', meaning serf or slave. The Robotics Industries Association have defined an industrial robot as 'a programmable, multifunctional manipulator designed to move materials, parts, tools, or special devices through variable programmed motions for the performance of a variety of tasks'.

Fig. 1.39 *An industrial robo*

Robots have many benefits within the manufacturing industry:

1 They offer increased machine utilisation with reduced labour costs.

2 They eliminate the need for people to undertake boring, repetitive tasks over long periods of time and they can operate in hazardous and 'unfriendly' conditions.

3 They are more accurate, less likely to make mistakes, stronger, and have a greater reach than humans.

4 Modern computer technology enables robots to be quickly reprogrammed, making them flexible, adaptable and able to perform many varied tasks.

Robot 'anatomy'

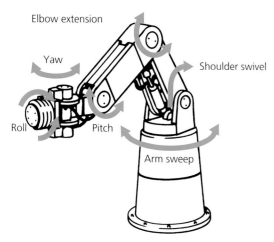

Fig. 1.40

The main feature of most industrial robots is the mechanical arm, which is designed to reproduce human-like movements in two or three dimensions. Its component parts even have names from human anatomy – arm, wrist, shoulder and elbow (Figure 1.40). Most robots incorporate one of four basic limb configurations.

Fig. 1.41 *Robot limb configurations.*

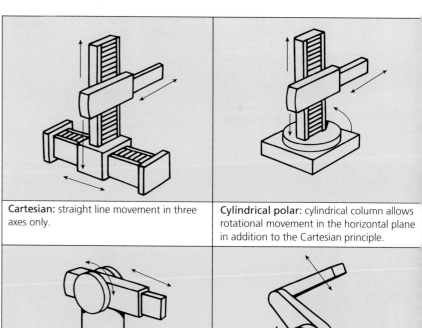

Cartesian: straight line movement in three axes only.

Cylindrical polar: cylindrical column allows rotational movement in the horizontal plane in addition to the Cartesian principle.

Spherical polar: arm rotates about a horizontal axis and can be extended along a linear path, column rotates about a vertical axis.

Articulated: known as anthropometric robots since they most closely simulate human arm movement.

The SCARA robot (Figure 1.42) is unique in that it does not have a separate wrist component but has rotational movement in the horizontal plane at two points, plus vertical movement of the end effector. They are sometimes referred to as 'pick and place' robots and are particularly popular for assembly operations, especially where components and parts are inserted, such as in the assembly of electronic circuit boards.

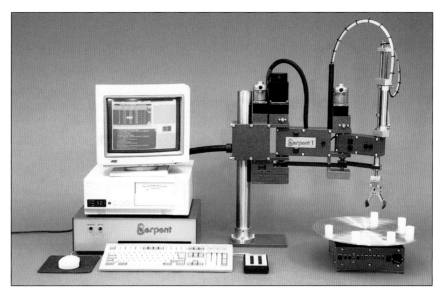

Fig. 1.42 *A SCARA robot and the computer that is controlling it*

Industrial robots comprise four basic systems:

1 The **mechanical structure** as described above.

2 The **drive system** which may be electrical (d.c. servo-motors or stepper motors), pneumatic, hydraulic or a combination of each.

3 The **tooling** (end effector) which will depend upon the task the robot is expected to perform. The most common means for materials handling is the mechanical gripper in which the component is held between one or more sets of mechanical fingers usually shaped to match the form of the component. Vacuum grippers are used for flat objects, magnetised grippers for holding ferrous components and adhesive grippers for handling fabrics.

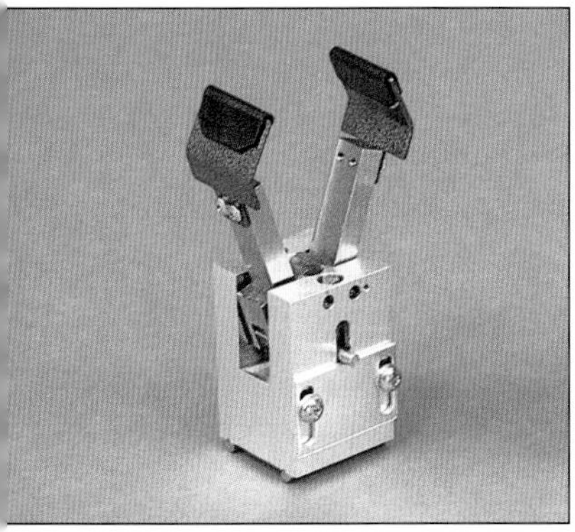

Fig. 1.43 *A mechanical gripper*

4 The **controller**, which is most often a microprocessor-based system although some still employ pneumatic or hydraulic logic control. Closed loop systems provide the feedback signals from the joints and end effectors to ensure that there is continuous accuracy and repeatability of movement and satisfactory completion of the task. Advanced robot controllers can interact with other machines in an integrated manufacturing set-up, avoid clashes, make decisions when things go wrong, provide data and respond to sensory inputs such as machine vision.

Robot programming

There are three ways of programming or teaching robots a sequence of movements:

1 **Walk-through** (nose-led): the operator manually moves the arm of the robot through the required sequence. The robot is able to retain the data for each stage of the movement cycle in its controller memory and will then repeat the learned movements as required.

2 **Lead-through:** this is a similar method to walk-through but the movements are initiated from a keyboard, joystick, simulator or control box.

3 **Off-line** (remote computer link): an important aspect of FMS is the ability to program robots from remote computer workstations in the same way as CNC machines receive instruction by Direct/Distributed Numerical Control. Robot actions are planned and simulated using sophisticated 2D and 3D animated graphic software at design, development and production planning stages of product and process planning. This ensures optimum use and efficiency.

Fig. 1.44 *A joystick control, used for lead-through programming*

MANUFACTURING SYSTEMS

There has been a radical change in recent years in the way British manufacturing has been organised and managed. Increased competition from other countries and the influence of work practices pioneered in the USA and Japan have been the motivation and the models for change. In addition the drive towards automation and computerisation has continued at a pace.

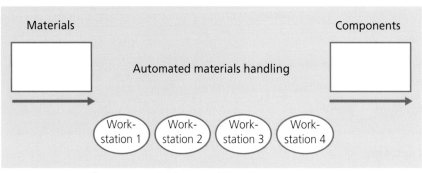

Fig. **1.45** *Diagram of an automated in-line manufacturing cell*

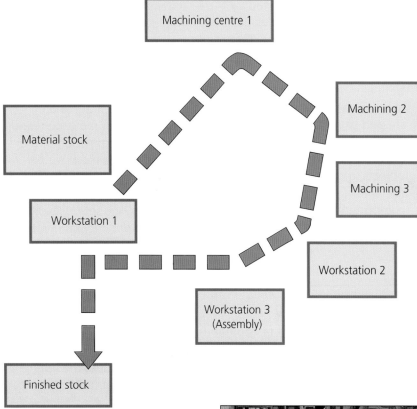

Fig. **1.46** *Diagram of a U-shaped manufacturing cell*

Cell production systems

Cellular manufacturing is becoming quite widespread in British industry. Production cells usually consist of a number of workstations grouped together to produce a single component or a number of similar components. A workstation may be a machining process, a manual operation such as assembly, or an inspection process. Each workstation performs a different operation on the component being made. Handling of components can be manual or automatic or a combination of both, and a cell will usually have a cell manager and a team of operators. Fully automated cells may have just one operator and a manager overseeing several such cells. The team is responsible for every aspect of production including quality control and maintenance.

Production cells can be designed in different ways. Two types are shown here in diagrammatic form. The automated in-line manufacturing cell (Figure 1.45) is best suited to high volume production. It is not flexible, however, and changes in product design can be difficult to accommodate.

Production cells like that shown in Figure 1.46 are used for medium volume production and are usually flexible enough to handle several similar products. The 'U' formation is a familiar one in manufacturing companies of all sizes. The workstations are relatively close together and the whole cell takes up the minimum amount of floor space.

Fig. **1.47** *In-line cell manufacturing wheel components*

Fig. 1.48 *An automated U-shaped manufacturing cell carrrying out a variety of machining processes on tractor transmission systems*

The advantages of cell production are:

- control of the cell is easier to manage
- production flow is easier to maintain
- problems are easily spotted and can be dealt with quickly
- control of work in progress (WIP) is greater
- communication is made easier
- more meaningful measure of team performance (i.e. productivity) is made
- targets and goals for the team are easier to set
- team motivation is improved
- components travel shorter distances, saving time and energy
- the team is more flexible to changing demands
- quality is more easily monitored and assured

The photographs in Figures 1.49 and 1.50 show manufacturing industry in the past. At Morris Motors in 1926, all cars were built upon a rigid chassis which was moved from one workstation to the next along a track mounted in the factory floor. Each item of assembly workers carried out their own particular contribution to the assembly before the car moved along to the next team. You can see how labour intensive the process was, with around four men at each workstation.

Fig. 1.49 *The assembly lines at Morris Motors' Cowley factory in 1926*

By the 1950s little had changed. In Figure 1.50, there is little evidence of automated processes. The man in the foreground is marking out using dividers, and you can see a hammer and centre punch for marking hole positions.

The modern production systems that we know today are releatively recent developments that have come about over the last thirty years with the development of electronics, and in particular, computer technology. We are now experiencing an ever-increasing speed of change, with people having to change and adapt their jobs, often several times, within their working lives.

Fig. 1.50 *Yardley's machine tool manufacturers, Liverpool, 1954*

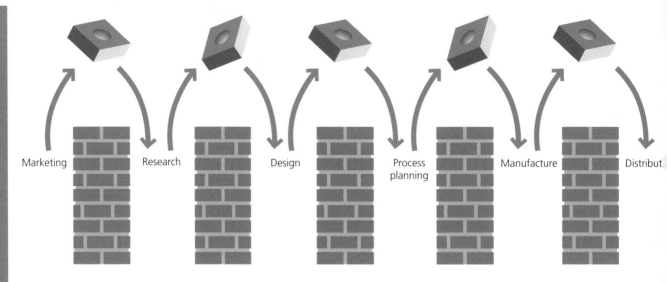

Fig. **1.51** *Over-the-wall engineering*

Concurrent engineering

We have already considered the important role of CADD in product design and development and its integration with CAM to produce more efficient manufacturing. These systems result in better overall performance, shorter lead times and greater control over quality. The key to this is the flow of information between the different departments, and its organisation and management.

Historically, British industry has been organised on a departmental basis with departments operating very much in isolation and preparing information only relevant to the next stage in the process. When the information – a design, a production plan or a schedule – was ready, it would be passed on to the next stage. This system has been compared to information being thrown over a brick wall and has been called 'over-the-wall engineering' because little communication between departments, other than those directly before and after in the flow of information, ever took place. When it did, developments were often well advanced, so alterations and amendments often proved costly and time consuming, and a considerable amount of paper was generated. Even when departments became computerised, the advantages in time were often lost because of the need to communicate with other departments by generating yet more paper.

Consider the flow of information in a typical manufacturing organisation (Figure 1.52). Computer technology has provided the means for effective information generation, storage and distribution. **Computer Aided Engineering (CAE)** links together design and draughting, modelling and analysis, production planning and control and manufacturing processes through an exchange of information that is current and accessible to them all. A concurrent engineering approach develops this by drawing together members from various departments on to the **product design team**, enabling the various specialists to interact and contribute without waiting for the previous person to finish. Incorrect interpretation of information can be quickly identified and corrected. Feedback is continuous since all the contributing departments and individuals have access to the same database. The main aim of concurrent engineering is to develop a product idea from customer requirement to finished item with the maximum efficiency.

Between	➡	And
Customer	➡	Marketing/Sales
Marketing/Sales	➡	Design office
Marketing/Sales	➡	Production planning
Design office	➡	Production planning
Production planning	➡	Production control
Production planning	➡	Quality control
Production control	➡	Quality control
Production control	➡	Manufacturing centres
Production control	➡	Stock control
Stock control	➡	Manufacturing centres
Manufacturing centres	➡	Quality control

Fig. **1.52** *Information flow*

The 'just in time' principle (JIT)

The traditional way to control manufacture has been to push the products through the production process in response to a plan or schedule according to the availability of components and resources. It has not always been related to true customer requirements and sometimes this has meant holding a stock of materials and resources in storage. This is expensive and stocks can take up valuable space, while finished stock may wait on shelves for some time before the next order comes in. When stocks are eventually needed they have to be transported from where they are stored to the manufacturing area, which takes more time. Also, changes in the design of a product could result in a significant amount of stock becoming redundant. The 'just in time' (JIT) philosophy prevents these situations from developing.

JIT was developed in Japan where manufacturers were interested in eliminating expensive stock as well as getting rid of waste in any form. They aimed at achieving what has become known as the 'five zeros'; zero stock, zero lead time (the time between customer order and product delivery), zero defects, zero breakdowns, and zero paperwork. Although these are almost impossible to achieve they are targets to be aimed at by looking continually for ways to make improvements. The just-in-time idea relies upon the timely delivery of materials, components and resources at the exact time they are needed for production to begin. In this way no large storage area is required, expensive stocks are eliminated and transport time becomes the responsibility of the supplier who may be from within the same factory or some distance away. The whole stock of materials, components and resources becomes **work in progress (WIP)**.

This system enables manufacturing companies to react very quickly to customer needs and changes in product design.

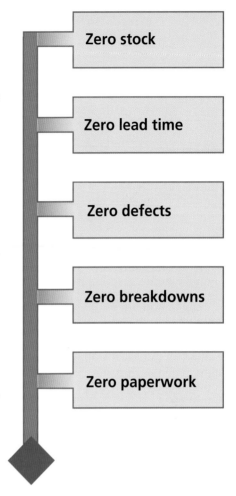

Fig. 1.53 *The five zeros*

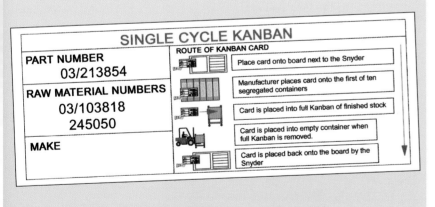

The Kanban system

'Kanban' is a word borrowed from the Japanese and means 'card' or 'sign'. It is a visual signal that enables the continuous flow of production in a JIT system to be controlled. A Kanban can be a card like the one shown in Figure 1.54, an empty container, or even a coloured ping-pong ball. It will usually carry information about the items being processed – name, part number, process and quantity being produced.

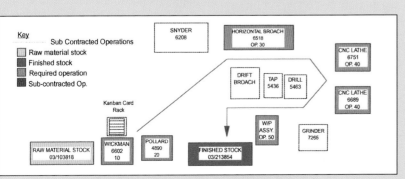

Fig. 1.54 *Both sides of a typical Kanban card*

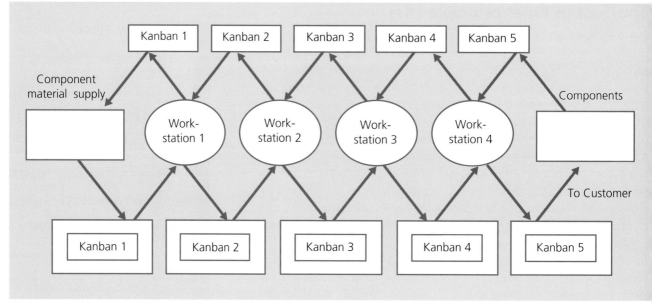

Fig. 1.55 *A Kanban system (using a card for each stage of the process)*

A Kanban can accompany a single item, batch or container depending on the type of component and product being manufactured. The Kanban is passed with the components from one workstation to the next. Look at the diagram above. When the same items are passed on further, the Kanban is replaced by a new one and the original is passed back to the previous workstation. There is always therefore, continuous communication between two adjacent workstations. By passing the Kanban back the workstation is saying 'send me some more components'. The diagram in Figure 1.55 illustrates how the system works. At the final stage of the production process (usually assembly and packaging) the product is prepared for despatch to the customer. As each completed product is taken for despatch a chain reaction is started back along the production line, because the corresponding Kanban is sent back each time to the previous workstation. In this way the products or components are said to be 'pulled through' the manufacturing process until the order is complete.

The Nagare system

This is a development of the Kanban system. It is characterised by:

- piece-by-piece work movement;
- no stock between workstations;
- one operator controlling a number of machines;
- manual transport, positioning and loading of parts on a number of machines;
- machines working automatically and ejecting the part at the end of each operation;
- the operator walking the line in process order;
- transportation of parts between machines being kept to a minimum.

The machines in this system are usually computer controlled so the operator only has to load and start each machine in turn, transferring each part from one machine to the next by hand.

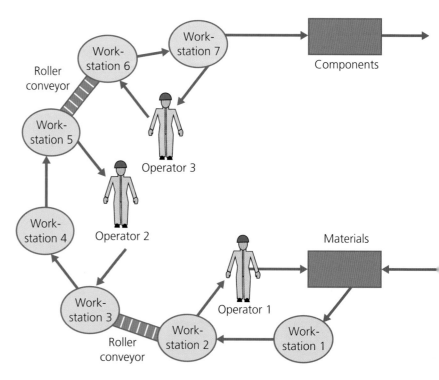

Fig. 1.56 *A Nagare system*

QUALITY ASSURANCE (QA)

British companies cannot be expected to compete internationally on the basis of cost alone since many other manufacturing nations, particularly those on the Pacific rim, have much lower labour costs. Whilst cost is a factor, quality of design, performance and customer service are increasingly the main reasons why one product is purchased in preference over another.

The notion of quality pervades the manufacturing industry. Quality assurance involves every part of the manufacturing organisation in the quest to ensure that the quality of its products in every way exceeds rather than just satisfies customer requirements. The **International Standard of Quality ISO 9000** (formerly BS5750) has been awarded to 50,000 British companies. It is, however, the Japanese who have led the way in pursuing ever better quality standards. '**Total quality management' (TQM)** and '**quality circles**' are two concepts that have been borrowed from the Japanese.

■ Total quality management is about establishing an attitude towards quality that permeates the whole of a company, not just its manufacturing sector. The responsibility for TQM lies within the management team who take the lead through example and through management structure and monitoring.

■ Quality circles are groups of production personnel who are encouraged to meet on a regular basis to sort out any issues that may have an adverse effect upon the quality of the final product. They are set up to encourage involvement and responsibility within a manufacturing company.

Quality control

Quality control is part of the quality assurance function. The '**right first time every time**' approach sets an ideal for the production unit that will not tolerate failure; agreed quality standards have to be met first time, every time. The quality and performance of products are monitored from beginning to end of the manufacturing process – from raw material to finished product. Inspection and testing are the practical ways in which quality control is applied.

Inspection

Inspection is the examination of the product and the materials from which it is made to determine if it meets the specified design standards. Factors such as dimensions, surface finish and appearance and in the case of materials, composition and structure, are considered.

Fig. 1.57 *The structure of a material is one of the many aspects that undergoes inspection – this is the microstructure of a steel sample*

Measurement is often essential but this can also be time consuming. An alternative to measuring, and one which is often simpler and faster, is to gauge the various characteristics of a product. The data generated by gauging is restricted to a simple pass/fail ratio. Measurement data is much more useful in keeping control of the process.

Fig. 1.58 *A turbine rotor being inspected using dial indicators to test for flatness and concentricity*

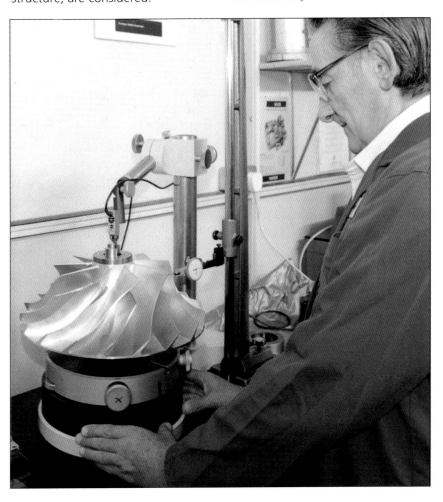

Testing

Testing is more often concerned with the functional aspects of a product. Does it work the way it should? Will it continue to work as it should over a period of time? Will it work in different environments? Sometimes the testing that has to be done is damaging or destructive, and only a limited number of products would be tested in this way because of the expense (as in Figure 1.59). Testing methods are constantly being developed to provide similar results without the need to damage or destroy. This is referred to as **'non-destructive testing' (NDT)**.

Tolerance

In practice, all the component parts of a product are manufactured to a level of 'tolerance' which is the acceptable degree by which they may vary from their '**nominal size**'. To try to obtain absolute precision where it is not needed would be a great waste of time and resources. Consider a simple wheel for a trolley or a barrow that is designed to run freely on a fixed axle. There must be some clearance between the wheel-bearing surface and the axle, otherwise the wheel would not be able to rotate. In Figure 1.61 the nominal size of the axle is 20mm and it must not be larger than this, but it is acceptable to be smaller by up to 0.3mm. Its size can then be said to be 20mm $^{+0.00}_{-0.30}$. Likewise, the wheel bearing must always be larger than the axle by at least 0.2mm. Its size can be said to be 20mm $^{+0.20}_{+0.50}$. On the drawing you can see these dimensions expressed in an alternative manner. It is essential to know that when a product reaches the assembly stage any components that are designed to fit together will do so without needing to be selective. In this example any wheels made to these tolerances will have a clearance fit with any axles made to the tolerance.

Fig. 1.59 *Testing a safe in fire*

Fig. 1.60 *An NDT test using an ultrasonic crack detection method*

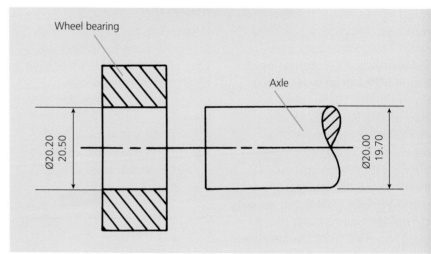

Fig. 1.61 *Dimensioned wheel bearing and axle*

Gauging

In order to determine whether or not a component is within tolerance limit, gauge inspection can be used. This is much quicker and less prone to error than measuring. Gauges are precision made instruments, sometimes pneumatic or electronic, but very often they can be as simple as the gap gauge and plug gauge shown in Figure 1.62.

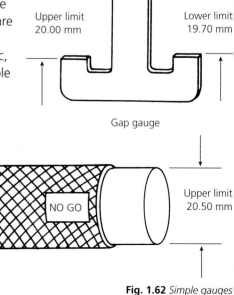

Fig. 1.62 *Simple gauges*

Statistical quality control

The results of inspection and testing, particularly when derived from samples of mass-produced products, are recorded on quality control charts (see Figure 1.63). The size and frequency of samples taken for inspection and testing depends upon the nature of the product and the scale of production. Quantities and times will be determined in advance by the application of mathematical formulas to ensure that representative samples are inspected and tested at intervals that are appropriate.

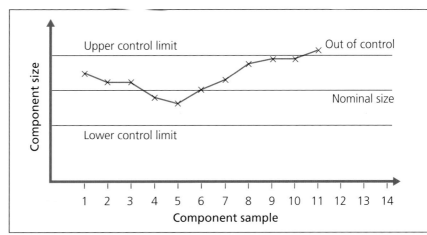

Fig. 1.63 *Quality control chart for a turned component, showing the size variation of samples inspected by a computer-controlled gauging machine*

Although the pursuance of quality aims for zero defects there is no process that will achieve this ideal consistently. '**Process capability**' is the term used to describe the variations inherent in any manufacturing process which, in normal circumstances can be predicted to fall within + or – three standard deviations of the mean, 99.73% of the time. The design tolerances in the product specification must be compatible with the process capability. Manufacturing variations can be attributed to a variety of sources: human errors, variations in the materials and components used,

and environmental conditions. The moment a sample goes 'out of control' (i.e. outside of the defined acceptable tolerance) then action has to be taken to remedy the problem. As with most other aspects of manufacturing the inspection and testing of products, components and parts, and the generation and presentation of the resulting data, is now often performed by computer-controlled machines.

Automation makes 100% inspection possible even in high volume production. By integrating inspection into the manufacturing process using computer technology the data being

generated at the point of inspection can be immediately 'fed back' to the process, and components and parts can be sorted and progressed according to quality levels, so a product may be categorised as acceptable, re-claimable, or scrap. The actual inspection is typically carried out by sensors that are controlled by and communicate with a computer. The data is usually digital in format. Sensors fall into two broad categories – those that make contact with the object being inspected and those that are able to measure or gauge the desired characteristics at a distance from the object. Coordinate measuring machines (Figure 1.65) are usually found 'off-line'. They employ a probe which has the ability to move in three-dimensional space providing a precise record of dimensional and geometrical features, such as hole locations and diameters, linear dimensions, sphere centre and diameter, flatness, and angular measurement between two planes.

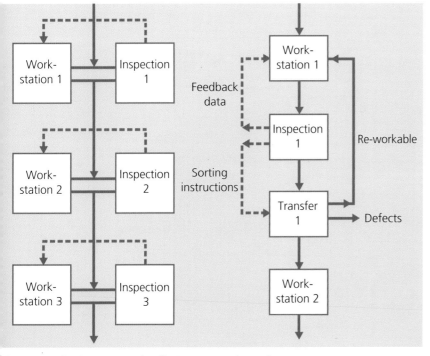

Fig. 1.64 *On-line/in process and on-line/post process inspection*

Fig. 1.65 *A coordinate measuring machine in use*

MANUFACTURING AND THE ENVIRONMENT

In recent years the manufacturing industry's effect upon the environment has become the focus of considerable attention. The drain upon the world's **non-renewable resources**, the long-term impact of many manufacturing processes, the '**greenhouse effect**' and the waste created by a multitude of industrial activities have become priorities for action as governments and technologists, particularly in Western Europe, have began to focus upon 'green' issues.

Inevitably computer technology has a positive role to play but brings with it other problems that directly affect the communities that have traditionally hosted manufacturing industries.

The ever increasing demand for higher productivity in the manufacturing industry has resulted in the growth of computerised, automated production systems. The resulting productivity gains have been offset by the loss of jobs. The effect upon individuals and local communities has been considerable. Whilst securing work for some in the short term it has also resulted in widespread, long-term unemployment for many others. As a result, a wealth of traditional manufacturing skills and knowledge, incompatible with computer technology, have been lost.

Fig. 1.66 *The manufacturing industry's effect on the environment is a cause for concern*

Waste

The twentieth century, and particularly the period since 1945 has seen a dramatic increase in the production of consumer products and of waste. Oil consumption in the period 1960–70 in developed economies was equal to the total oil produced before 1960. Coal consumption since 1940 has exceeded all the coal used in the previous 900 years. The richest 25% of the world's population use 80% of global energy, consume 85% of global chemical production and 90% of global automobile production. The increase in waste has proceeded unchecked. The manufacture of products that do not need to be replaced are often not considered good for business. Durability decreases purchasing, replacement and repairs, but is advantageous environmentally. Fashion considerations are yet another factor encouraging the 'throw-away society' to manufacture and generate waste.

Fig. 1.67 *Total waste production in the UK (Dept of the Environment, 1995)*

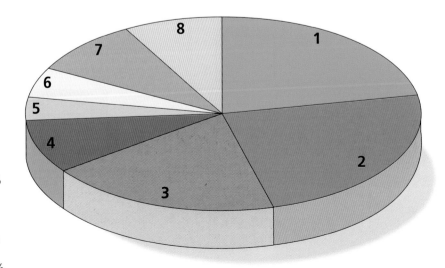

1 Agriculture **22%**
2 Minerals **24%**
3 Industrial **19%**
4 Sewerage **9%**
5 Commercial **4%**
6 Household **5%**
7 Construction **9%**
8 Dredged soil **8%**

Much of the solid waste generated is disposed of either through landfill or incineration. Landfill usually results in the emission of landfill gases, which on entering the atmosphere contribute to the greenhouse effect. Methane gas from landfill sites is thought to be up to 60 times more powerful than carbon dioxide in its contribution to global warming so sites have to be carefully engineered with methane being collected and allowed to 'flare off'. Some liquid waste also ends up in landfill sites and great care has to be taken to avoid such substances leaching into surrounding land and water courses. Other liquid wastes can be disposed of into rivers at predetermined levels, although ensuring these levels are kept to is a very difficult task. Almost all manufacturing industries use, and have to dispose of, chemicals at some point in the production process.

Incineration can result in residual ash and the emission of combustion gases that may contain toxic dioxins and furans produced from the burning of chlorine-containing compounds such as plastics and bleached paper. The control of such emissions is monitored by government agencies and plants have to reach the highest standards for the filtration of such pollutants. New technology has resulted in the recovery of heat and electrical energy from incineration plants, although Britain still lags behind major industrial competitors in the amount of energy recovered from waste.

Fig. 1.68 *Industrial effluent*

Fig. 1.71 *Inspectors monitoring pollution levels*

Fig. 1.70 *An artist's impression of the 'waste tyres-to-energy' incineration plant in Wolverhampton*

In an attempt to reduce the effect of industrial waste, much work is being done to find ways of turning waste from one process into the raw materials for others. For example, residues from the production of PVC are used in the production of cleaning agents. This is, however, a new field and it is proving very difficult to produce cost effective, pure, contaminate-free recycled materials.

The traditional view of the manufacturing industry has been of grimy chimneys belching out sulphurous, choking smoke. The Control of Pollution Act 1974 made it illegal to emit dark smoke, fumes and excessive amounts of dust and grit, and The Environmental Protection Act 1990 and the Environment Act 1995 placed even more legal requirements upon industry to ensure more protection for the environment. Legislation of this type, however, always has a price and has resulted in increased manufacturing costs that are passed on through the industry, eventually reaching the consumer.

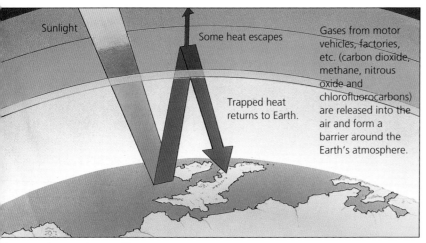

Sunlight

Some heat escapes

Gases from motor vehicles, factories, etc. (carbon dioxide, methane, nitrous oxide and chlorofluorocarbons) are released into the air and form a barrier around the Earth's atmosphere.

Trapped heat returns to Earth.

Fig. 1.69 *The greenhouse effect*

Life cycle analysis (LCA)

Life cycle analysis is a recent development aimed at reducing waste from manufactured products and the processes used to make them. LCA involves making detailed measurements during the manufacture of the product, from the extraction and distribution of the raw materials used in its production, through its manufacture, use, possible recycling and eventual disposal. Manufacturers are able to quantify how much energy and raw materials are used, and how much solid, liquid and gaseous waste is generated at each stage of the product's life cycle. The resulting data is an indication to the manufacturer of the impact their product will have upon the environment. Such information can be utilised by the design team to develop products that are more environmentally friendly and less demanding upon finite resources and energy. In an environmentally conscious world manufacturers are beginning to recognise that products designed on the basis of LCA will have a competitive edge.

Life cycle analysis is providing the foundation for Britain's 'national waste strategy' which will set standards as targets to ensure that waste is managed in a 'sustainable' way. The Government's policy for achieving this is set out in the form of a 'waste hierarchy', with the most important strategy first and the least attractive last (Figure 1.73).

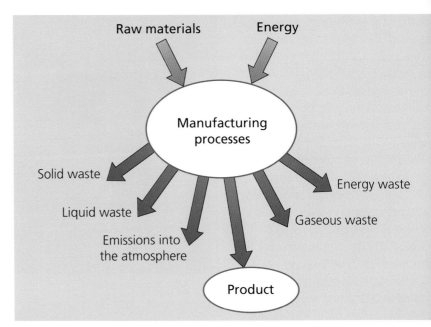

Fig. 1.72 *Life cycle analysis*

Recycling

Recycling of 'waste' materials and 'waste' energy wherever possible is vital for the long-term prosperity of all the earth's inhabitants. Recycling can be extremely energy efficient if the waste is readily available. It is often the collection and transportation of recyclable waste that makes the whole process non-viable. For example, it is 20 times as efficient, in terms of energy usage, to recycle aluminium rather than to produce it from raw bauxite ore. However, if transportation and collection costs increase and energy costs fall, then in terms of pure economics it becomes less worthwhile to recycle.

Fortunately, increasingly more and more industries within the developed industrial countries have a responsible attitude towards this problem and prefer to recycle waste. Many products, particularly those made

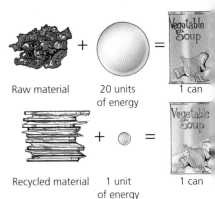

Fig. 1.74 *Recycling*

from wood pulp, such as paper and card, are promoted on the fact that they are made from recycled raw materials. Recycling can be good for business. Governments can also help by offering incentives through tax and legislation to encourage recycling. Industry in poorer countries, however, is less likely to be able to look further than the short-term gains that may enable them to be competitive with their richer neighbours. To add to this dilemma many countries within the developing world have ready access to minerals, timber and fossil fuels that are consumed in response to the increasing demand for manufactured consumer goods. The emphasis on recycling is usually on materials in household waste, but many other options exist. In Britain and other densely populated industrial nations land previously used for industry has been recovered to re-use it for housing, recreation and new industry.

Fig. 1.73 *A waste strategy*

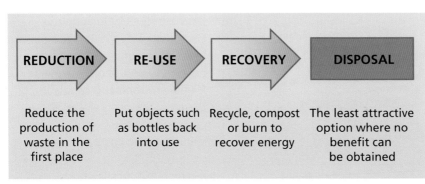

ig. 1.75 *The Merry Hill shopping complex in Dudley was once the site of a steelworks*

There are five clear potential points for recycling worth considering:

Waste minimisation and prevention is best dealt with in manufacturing production processes, where it is estimated that 90% of pollution originates (i.e. do not create the waste in the first place).

Pre-consumer recycling and re-using unavoidable waste within the production process is easier and more economical for industry than attempting to recycle materials that have been in the hands of the consumer.

3 Product re-use by reclamation and/or repair prolongs a product's usefulness, such as using returnable drinks containers or re-treading car tyres.

4 Primary recovery is the most familiar recycling of waste that produces new raw materials. It is appropriate for recycling used paper, card, fabrics, metals and many types of plastics.

Recycling waste:
- conserves non-renewable resources and the sites they are taken from;
- reduces energy consumption and greenhouse gas emissions;
- controls the pollution involved in the manufacturing process, the extractive industries and the disposal of waste;
- reduces the dependence on imported raw materials;
- creates work in areas of high unemployment through community-based industries such as paper and glass making, renovation and repair of consumer products;
- encourages the environmental educational benefits of participation in recycling.

5 Secondary recovery is the reclaiming of energy from waste, such as electrical energy from incineration plants and includes new processes such as high temperature burning of tyres and plastics and the recovery of landfill gas as a fuel source.

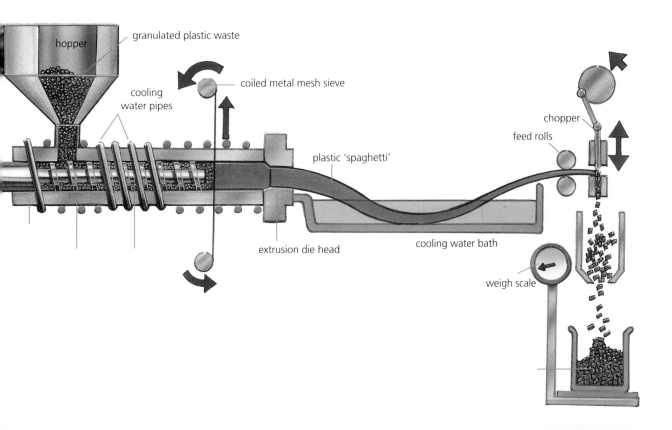

Putting it into practice

1. Consider a job or manufacturing process from the time of the industrial revolution that still exists today. Say how the process has evolved and changed over the last 200 years?

2. a) What do you understand by the term 'the country's infrastructureí?
b) What effect does the infrastructure of a country have upon its ability to develop as an industrial nation?

3. What is meant by the term 'service industry', and what is the effect upon the economy of a country if very large numbers of people are employed in service industries.

4. What function does 'market research' play in manufacturing industries? Suggest some ways that market research might be carried out.

5. Give some examples where developments in technology have 'pushed' product development forwards.

6. Carry out a detailed product analysis of a small product, such as a stapler or a paper punch. Make sketch drawings of all of the component parts and say what materials they are made from, the processes that have been used in their manufacture and the order in which the product has been assembled.

7. a) What is the difference between 'computer aided design' and 'computer aided draughting'.
b) Draw up a table that shows the advantages and the disadvantages of installing and using CADD systems.

8. Trace the complete cycle of manufacture of a product that is manufactured near to your school or where you live. Make reference to all of the 'Process Operations' identified on page 12 of this chapter.

9. Draw up a table of abbreviations, such as CIM, that are associated with manufacturing. State clearly what each one stands for.

10. What is meant by the following CNC related terms;
i) binary numbers, **ii)** part programming,
iii) direct numerical control, **iv)** command words,
v) miscellaneous functions ('M' codes), **vi)** axes of movement, **vii)** 3D axes control, **viii)** zero datum point,
ix) absolute coordinate system, **x)** CADCAM.

11. Stock control and materials handling are important features of any manufacturing system.
a) Why are they considered to be so important?
b) Compare materials handling within a FMS system to that within a traditional structure.
c) What is meant by 'Just in Time'?

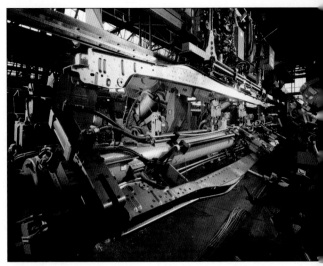

Fig. 1.77 *Robot spot welding*

12. a) Illustrate four different types of robot and suggest an application of each type.
b) How are robots programmed to perform the task required?
c) Outline some of the advantages and the disadvantages of robots within manufacturing industry.

13. Why is 'cell production' becoming the most favoured arrangement in most large manufacturing industries?

14. How does 'concurrent engineering' speed up the time it takes to develop a new product and get it onto the market?

15. Describe in detail the various ways that quality can be developed and 'assured' within manufacturing industry.

16. Why is it necessary for dimensions to have a 'tolerance'?

17. What part does 'inspection' play in quality assurance and how does gauging and testing fit into this process?

18. What is meant by the terms; **i)** 'non-renewable resources' **ii)** 'the greenhouse effect'?

19. a) Why has the concern for manufacturing industry to be more aware of the waste it creates increased over recent years?
b) What steps are being taken to ensure that waste creation and pollution are being controlled?

20. Recycling waste materials is desirable but it is not always economically sound practice. How can consumers and manufacturers be encouraged to recycle more of their waste?

2. Investigating materials

Since the dawn of civilisation, human existence has always been closely tied to natural materials. Bone, stone and wood were first used to make tools and weapons that would provide the means of obtaining food, clothing and shelter. Later, the use of other natural materials such as minerals and ore gradually led to new ones being introduced – bronze, for instance, was the result of alloying (mixing) several minerals together. Today, materials left behind by settlers through the ages can still be used by historians as references when tracing and describing the early stages of development in the Stone, Bronze and Iron Age.

New materials and processes evolved only gradually through the centuries, until the Industrial Revolution heralded in a period of rapid and dramatic change. Increasing scientific knowledge led the development of new materials away from a 'trial and error' approach to a science-based technological approach. This led to new methods of working and to the introduction of new materials. Soon there was an enormous growth in the use and processing of natural resources. By the end of the nineteenth century, the impact of technology could be clearly seen in the new forms and designs for cast and wrought iron incorporated in bridges, buildings and means of transport.

The twentieth century brought about other significant changes, at an even faster pace. The widespread use of reinforced concrete changed the environment. It was used for roads, better bridges and high-rise buildings.

It is plastic, however, that is the material phenomenon of the age. This synthetic manufactured material affects every aspect of modern life, from toiletries and food-packaging to our buildings and transport. Its use helps to maintain our comfortable living standards and continues to improve our quality of life.

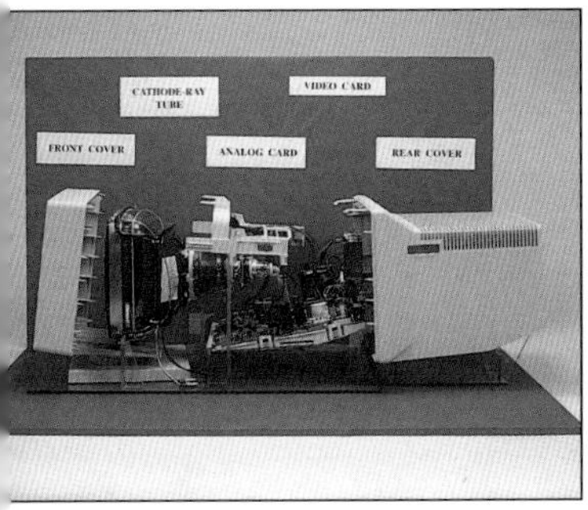

Fig. 2.1 *Many nineteenth-century structures were of cast iron and glass*

Fig. 2.2 *Canary Wharf tower – a high-rise structure*

Fig. 2.3 *Plastics act as insulators in a computer*

Today, design and technology have a large part to play in the making of products from materials that can easily be re-cycled. Our 'throw-away' age is gradually awakening to the need to become 'environmentally friendly' and to conserve the Earth's non-renewable resources. It is estimated that the average family of four throws away the equivalent of six trees, 50 kilograms of metal and 40 kilograms of plastic every year. As well as the necessity to stop pollution by waste material, there is also a need to reduce costs and to conserve energy and resources for the benefit of future generations.

This chapter investigates and identifies the different types of material available to enable you to make informed decisions concerning the materials to use in your project work.

Fig. 2.4 *The green parts of this BMW can be recycled, the blue parts are made from recycled material*

Fig. 2.5 *A metal-formed structure for a roller coaster*

CHOOSING MATERIALS

In today's world the problem of choosing suitable materials has become more complex because of the possibilities offered by an ever-expanding resource of available materials. To solve the problem of which material is to be used for a particular item, several factors must be considered.

Materials used in products must be perfectly suited to the **function** of the finished product. Requirements such as hardness, rigidity, weight, flexibility, colour or texture may be needed for an item to perform its function properly. These requirements must be taken into account and matched against the properties and characteristics of suitable materials, which are often in varied combinations and forms.

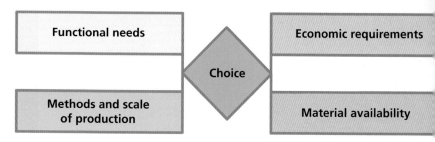

Fig. 2.7 *Choosing materials*

Fig. 2.6 *Materials are available in a wide range of forms*

Methods and scale of production

The interrelationship between the material and the method (means) and scale of manufacture of the product must be considered if quality and high standards of precision are to be achieved while maintaining economy. An awareness of the scale of production, whether a one-off, a batch, or mass production is

important. Materials must also be assessed for their compatibility and suitability to the industrial processes and techniques used in manufacturing specific components, for example some materials that can change state may be suitable for casting and deforming. The material chosen is likely to influence the method of fabrication (making) and type of finish that is applied.

Availability

Most types of material such as wire, sheet, plate, flat strip, round and square bar, tube, angle, extruded section, granular chips, pellets and viscous fluids (see Figure 2.6) are available from suppliers in standard forms, shapes and sizes. Standardisation affects both the quality and size of materials, and traditional practice and demand have established **standard preferred sizes**. Using non-standard materials in manufacturing can increase costs considerably. Specialist information, available from suppliers' catalogues, is not always presented in a simple manner and often needs careful research. It is essential to check the availability with stockists and local suppliers to ensure your material specifications can be met.

Economic requirements

Correct costing is vital in the manufacture of any product, not only in the purchasing of materials, but also in the working and processing of them, such as in machining, joining and finishing. Any likely wastage must be included in the costing. In industry, time and labour costs also affect the overall viability of production. Choice of material can help to determine quality and cost, and often the use of expensive materials can be justified by the low cost of processing them. Some of the variable factors are illustrated in Figure 2.8 in the range of materials used to provide drinking vessels. They include wine glassware, china cups, pottery mugs, silver goblets, pewter tankards, plastic beakers and waxed paper or plastic cups. They range from traditional manufacture to volume mass-production, with some of the material solutions matching the convenience requirements of our disposable, 'throw-away' age.

Combining materials

It is only when all the information is collated (collected together) that decisions can be made. Use of a simple computer spreadsheet can help to determine comparative costs, but the final choice is often a compromise between function and cost. Usually several alternatives, including combinations of materials, are available. There is rarely only one correct answer or best material.

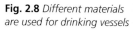

Fig. 2.8 *Different materials are used for drinking vessels*

Physical properties

■ **Fusibility** is an essential feature of processes like casting, moulding, soldering and welding. Metals and some plastics change into a liquid (molten) state when heated to a certain temperature. Known as the **melting point**, this temperature varies considerably between materials (see Figure 2.9).

■ **Density** is the amount of matter in an object (mass) per unit of space occupied by the matter (volume). Relative density is the ratio of the density of a substance with that of pure water at a temperature of 4°C.

Fig. 2.9 *Melting points*

■ **Electrical conductivity** is the movement of electricity through a substance. Some materials, especially metals such as gold, silver and copper, liquids (electrolytes) and some gases are good conductors, allowing electricity to pass through them easily. Some materials resist the flow of electric current, and those that offer high resistance to electrical flow are called insulators. These include non-metals, such as wood, ceramics and mica, glass and many plastics, PVC, and nylon. Semi-

Fig. 2.10a *Ceramic insulators on a pylon*

Fig 2.10b *Plastic insulates a silicon chip*

conductors form the basis of electric components and allow current to flow only under certain conditions. Made from silicon and germanium these semi-conductors, like those shown in Figure 2.10, are poor conductors in their pure state but the addition of impurities alters their electrical resistance.

Fig 2.11 *Fire resistant materials are needed on a space shuttle*

■ **Thermal conductivity** is the movement of heat through a substance. Conduction rates vary according to a material's reaction to heat and it is important that the right materials are chosen for the specific task. For instance, spacecraft (Figure 2.11) endure very low temperatures in outer space, but on re-entering the Earth's atmosphere, must resist extremely high temperatures. Fire resistant materials like asbestos and silica tiles are used to prevent vaporisation. **Thermal insulators**, generally non-metals, have low-value conductivity. They are used to prevent heat gain or losses, and are often materials such as fire resistant plastics used for pan handles. Air is one of the best

insulators and is used with the other materials, such as those illustrated in Figure 2.12, to reduce heat loss in the home. **Thermal expansion** relates to linear expansion, which is a material's fractional change in length, expanding when hot and shrinking upon cooling.

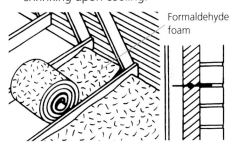

Formaldehyde foam

Fig 2.12 *Loft and cavity wall insulation*

■ **Optical properties** of materials vary according to how materials react to light and heat. Some materials react to light and heat by bouncing it back. This is called reflection (Figure 2.13). Some materials radiate (emit) light, heat or other electromagnetic waves and others absorb them. Materials may be **transparent** (see-through), **translucent** (allowing some light through) or opaque (not allowing light through). **Colour** is another significant factor when it comes to choosing a material. It can be a means of identification and can also determine a material's suitability for decorative work, and it can be modified and changed.

Fig 2.13 *Materials that reflect light are needed for a car*

Mechanical properties

The mechanical properties of materials are associated with how a material reacts to the application of forces. When a force is of sufficient strength, **deformation** of the material will occur. This may be either temporary (**elastic**) or permanent (**plastic**) which enables components to be easily shaped and produced cheaply, using less energy.

FORCE per unit is called STRESS

Displacement (distortion by extension or compression) per unit length is called STRAIN

$$Stress = \frac{Load}{Area}$$

$$Strain = \frac{Extension}{Overall\ length}$$

Fig. 2.14 *Force, stress and strain*

■ **Strength** is a material's ability to withstand force without breaking or permanently bending. Different types of strength resist different forces.

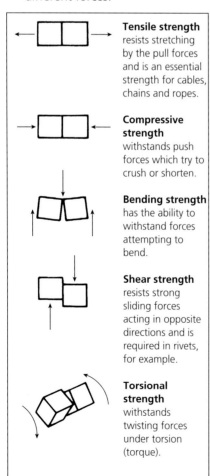

Tensile strength resists stretching by the pull forces and is an essential strength for cables, chains and ropes.

Compressive strength withstands push forces which try to crush or shorten.

Bending strength has the ability to withstand forces attempting to bend.

Shear strength resists strong sliding forces acting in opposite directions and is required in rivets, for example.

Torsional strength withstands twisting forces under torsion (torque).

Fig. 2.15 *Types of strength*

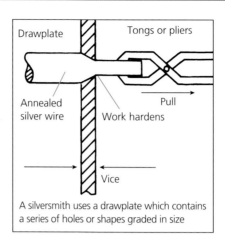

A silversmith uses a drawplate which contains a series of holes or shapes graded in size

Fig. 2.16 *Reducing jewellery wire by pulling through a die*

■ **Elasticity** is the ability to flex and bend when subjected to forces and regain normal shape and size when loads are removed, such as in an elastic band. Most structures need to possess this in some degree.

■ **Plasticity** is the ability to be permanently changed in shape, by external blows or pressure, without cracking or breaking. Some materials are more plastic when heated. Two other terms are associated with plasticity. The first is **malleability** – the extent to which materials can undergo permanent deformation in all directions under compression, such as hammering, pressing or rolling, without rupturing or cracking. The second is **ductility** – the ability to undergo cold plastic deformation by bending, twisting or stretching. Figure 2.16 shows how a permanent reduction in a cross-section of wire jewellery is achieved by pulling it through a die without rupturing. All ductile materials are malleable, but malleable materials are not necessarily ductile.

Order	Malleability	Ductility
1	Silver	Silver
2	Copper	Iron
3	Aluminium	Nickel
4	Tin	Copper
5	Lead	Aluminium
6	Zinc	Zinc
7	Iron	Tin
8	Nickel	Lead

Fig. 2.17 *Malleability and ductility of metals*

■ **Hardness** is a complex property. I is the ability to resist abrasive wear and indentation or deformation. It an important quality of cutting too such as drills, files and saws. Abrasives also depend upon hardness to be effective.

■ **Toughness** is the ability to withstand the sudden forces (stresses) of shocks or blows witho fracture, as well as resist cracking when subjected to bending and shear loads. **Brittleness** is the opposite, showing little or no strai (plastic deformation) before fracture. Brittle materials do not allow pressing or bending, but simply break. Repeated and revers stresses, and bending and tension forces wear down the toughness o some materials causing fatigue (break down).

■ **Durability** is the ability to withsta wear and tear and deterioration b weathering. Changes may result in mechanical weakening as well as i appearance. **Corrosion** is the chemical attack of the surface of a material and is common in metals. However, plastics generally are les prone to decay, which has led to their widespread use.

■ **Stability** is the resistance to changes in shape and size. Timbe tends to 'warp' and twist with changes in humidity. Metals and some plastics gradually deform when subjected to stress for long periods. The gradual extension of material under a load is known as 'creep'. Materials used in produc that have to withstand high temperatures and rotational speed such as turbine blades (Figure 2.1 must be creep resistant.

Fig. 2.18 *Gas turbine blades*

MECHANICAL TESTS

Standardised testing, by deformation or destruction, is carried out to determine the mechanical properties of materials. Simple graphs can be used to plot and present results.

Tensile tests

To measure strength and indicate ductility, a prepared sample test piece (Figure 2.19) is gripped in the jaws of the testing machine and an increasing force applied.

The graph (Figure 2.19) shows that when a small force is applied, extension is proportional to the force. If the force is removed before point 'A' is reached, the material will return to its original shape and size (elastic). However when the elastic limit is reached, a small permanent extension will remain (plastic). Further force causes the sample to stretch rapidly. **Necking** occurs shortly after the maximum force 'M' has been reached and just before final fracture occurs. If a nominal stress/strain curve is drawn from information recorded in the test, **Young's Modulus of Elasticity** 'E' i.e. the slope of the straight portion O-A) can be determined.

$$E = \frac{\text{Increase in stress}}{\text{Increase in strain}}$$

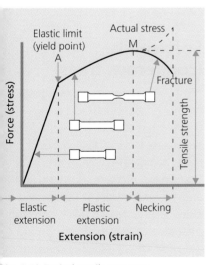

Fig. 2.19 *Typical tensile test curve*

Hardness testing

■ **Scratch tests**, like the Moh's scale used by mineralogists, rank a number of substances in an order whereby each will scratch the one coming before it. A diamond has a hardness rating of 10. This can be duplicated by scratching small samples with a scriber, a diamond (glass cutter), tungsten carbide (tip of masonry drill) or even abrasives.

■ **Indentation tests** are of different types, but all use the principle of forcing a hard object into the surface of the test specimen, with the hardness value based either upon the surface area of the impression, or depth of indentation (Figure 2.20). The **Brinell test** uses a hard steel ball as an indenter pressed under a constant load. The **Vickers test** uses a small pyramid-shaped diamond tool to make indentations. The two diagonals are measured and averaged to give the degree of hardness. The **Rockwell test** uses either a steel ball or diamond cone, but gives rapid results automatically, measuring the depth of indentation.

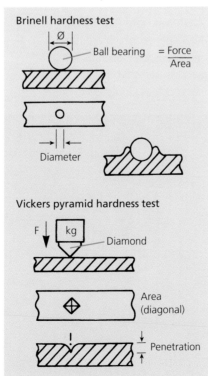

Fig. 2.20 *Indentation tests*

Toughness testing

■ The ability to withstand sudden shocks or blows is assessed in the **IZOD Impact test**. A notched specimen (Figure 2.21) is struck by a heavy pendulum, falling under gravity. The extreme swing position indicates strength. The tougher the material, the more energy is absorbed in breaking it and the smaller the extent of the swing after fracture.

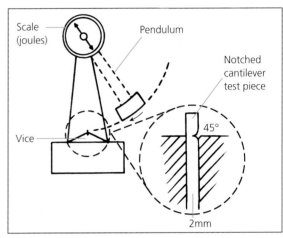

Fig. 2.21 *Impact tester*

■ **Photoelasticity** uses polarised light to analyse stress in structural pieces. Acrylic or epoxy resin test pieces, put under stress in a **polariscope** show the effect of the way light passes through (Figure 2.22). The scientific term is '**birefringence**', which means 'doubly defracted'.

Fig. 2.22 *Birefringence patterns in various plastic objects*

WOOD

Wood is an adaptable, versatile material. It has been used to meet human needs as fuel for fires, weapons for protection and hunting, tools, utensils and even footwear. Structural uses include housing, from crude shelters to sophisticated Japanese temples and the mass-produced timber-framed units of today. Many original oak beams and rafters (as shown in Figure 2.23) are still to be found in the roofs of old buildings.

The first machines, for example clocks and windmills, were made in wood. It has also seen a wide divergence of use in transportation – being used in carts, coaches, canoes, ships and aircraft. We probably most associate this material with furniture. The traditional Windsor chair (see Figure 2.24) illustrates the material's diverse nature: elm makes up the seat, ash the bent bows, arms and back and beech the turned legs and sticks. By contrast, the amount of solid wood used in today's furniture industry is small, in comparison with that of 'manufactured' timber.

Fig. 2.24 *Traditional Windsor chair*

Conditions of supply and demand continue to affect this important resource. Forests are one of the few natural resources capable of self-renewal within a short timescale. Britain is one of the least wooded countries in Europe and imports 90% of its timber needs.

Fig. 2.23 *Oak beams and rafters*

The tree not only has a unique beauty of its own, but provides timber peculiar to its own species. Characteristics include colour, grain pattern, texture, strength, weight, stability, durability and ease of working. Trees are highly developed plants, made up of minute cells composed of **cellulose**. The cells vary in number, form and function, enabling these cellular structures to be identified and distinguished from one another. One group of plants is called **endogens**, in which growth takes place inwardly in a hollow stem, such as bamboo, palms and tree ferns, and begins growth with a monocotyledon (single seed leaf). This group has little commercial value. **Exogens** (dicotyledons, i.e. with two seed leaves) however, are outward growing, increasing in size by adding new tissue in the form of concentric growth rings, each growing season. They divide into two classes: angiosperms which are broadleaf, deciduous trees such as ash; and gymnosperms which are coniferous trees such as Douglas fir (Figure 2.26).

Hardwoods (broadleaf trees)

Most broadleaf trees are deciduous (leaf losing) and have covered seeds, often enclosed in fruit or nuts such as apples or acorns. Although known as hardwoods, this is a botanical division and is not always a guide to the softness or hardness of the texture of the wood. Broadleaf trees grow in warm temperate climates in Europe, Japan, New Zealand, Chile and tropical regions of Central and South America, Africa and Asia. Growth is generally slow, taking around a hundred years to reach maturity, making them expensive to use commercially. Most hardwoods shed their leaves annually, but some, such as holly and laurel, retain their leaves for much longer and are known as evergreen. Tropical hardwoods also retain their leaves and consequently grow more quickly and to a greater scale.

Fig. 2.25 *English oak (hardwood, deciduous)*

Softwoods (coniferous trees)

Most softwood trees are evergreen (larch is an exception) and are coniferous (cone bearing) with thin, needle-like leaves. They grow in colder temperate climates in Scandinavia, Canada, Northern Russia and at high altitudes in Europe and elsewhere. Growth is quick, reaching maturity in about thirty years, making them relatively cheap to use commercially in manufacturing.

Fig. 2.26 *Scots pine (softwood, coniferous)*

Growth and structure of trees

The structure of a typical tree is shown in Figure 2.27. The **roots** have root hairs which absorb water and dissolve mineral salts to make crude sap, while others support and anchor the tree. **Leaves** take in carbon dioxide from the atmosphere and sunlight is absorbed by the **chlorophyll** (green pigment). The energy from the sun is used to synthesise organic compounds in a complex chemical reaction, known as **photosynthesis**. **Sapwood** is newly formed wood, often light in colour and quite soft, made up of **xylem** cells. Sap, water and mineral salts are carried by suction pressure through these cells up the tree to the leaves, where they are manufactured into food. The **heartwood** is matured sapwood made up of **lignified** (hardened) cells which give strength and support the tree. It provides the most commercially useful part of the tree, being much harder, stronger and often darker in colour. **Pith** (**medulla**) is to be found throughout the length of the tree, the remains of the earliest growth extending upwards. The **medullary rays** are thin sheets of tissue that extend from the cambium to the pith (medulla), like the spokes of a wheel. They conduct and distribute waste products horizontally for storage in mature cells, forming **figure** (silver grain in oak). **Bark** is the corky skin which prevents **transpiration** (water loss) from the trunk and serves to protect the tree against damage and extremes of temperature. The **bast** (phloem) is the inner bark made up of living tissue – **phloem cells** – which carry food downwards from the leaves to other parts of the tree. The **cambium layer** is where growth takes place and it completely surrounds the sapwood. New wood cells (xylem) are formed on the inside and to a lesser extent, new phloem cells on the outside. The **annual rings**, called growth rings, represent one season's growth and each band is made up of two distinct layers. The inner spring wood consists of large,

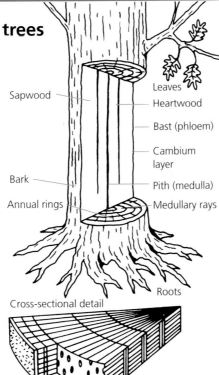

Fig. 2.27 The structure of a tree

soft, thin-walled cells, which help the sap to flow. In summer, with less sap, the cells are smaller, thicker walled and more dense. Viewed in cross-section, this growth cycle shows distinct bands, by which the age of the tree can easily be determined. Some tropical timbers show no visible annual rings because growth takes place uniformly throughout the year.

The cell structure of **softwood** is shown in Figure 2.28. The **tracheids**, which are thin, elongated tubes sealed at the ends and spliced together, form the bulk of the timber. Communication between the cells for the passage of sap and food takes place through **pits** which are small openings in the walls. The cells harden with age and serve to support the tree. They form in radial rows and it is the direction in which they lie that makes up the grain of the wood. **Parenchymas** are smaller, with simple type pits and make up the remaining cells. They include thin rays that are almost invisible. However, they are a reliable means of identification between species when magnified. The resin canals are the rays by which resins and gums are carried.

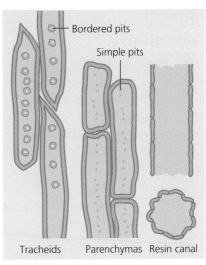

Fig. 2.28 Softwood cells

The cell structure of **hardwood** is shown in Figure 2.29. **Fibres** make up the bulk. Small and sharp they mechanically support the tree, but do not carry sap. **Vessels** (pores) provide a positive means of identification. They form ducts or tubes that extend the whole length of the tree carrying food and appear in two different forms. One is a **diffuse porous** form where they are evenly spread throughout the tree, as in beech, birch, sycamore and tropical hardwoods like ebony and mahogany. The other form is a **ring porous** form where the pores appear quite large in early spring growth and much smaller in summer growth as in ash, elm and oak. **Parenchymas** form radially in rays, which are often prominent, especially in oak, where they produce the familiar silver grain.

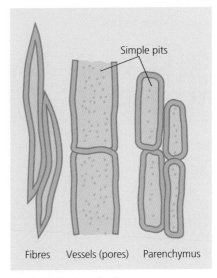

Fig. 2.29 Hardwood cells

Conversion and seasoning

Trees are felled when they reach maturity and felling normally takes place in winter, when there is less sap and moisture in the tree. The felled trees are collected together and stacked near a road or railway for transporting to a saw mill. In some countries, such as Canada, the spring tidal waters of flowing rivers are used to float logs downstream for processing.

'**Conversion**' is the term given to sawing the log into marketable timber. The log may be sawn in a way that promotes features like grain pattern and figure, as well as stability in use. Baulks of timber are sometimes produced, which involves removing the unwanted sapwood (Figure 2.31). Two basic methods are used for conversion. The **slab** method (also called **plain**, or **through-and-through sawn**) is shown in Figure 2.31. This is the simplest, quickest and cheapest and involves the log being cut into parallel slices (slabs) of variable thickness. The other method, which is called **quarter** or **radial sawn** (Figure 2.31) is more expensive, involving first sawing the log into quarters. In practice, several near-radial methods are used to reduce waste and simplify sawing. They produce excellent quality, stable timber which is less likely to be affected by 'movement' and warping (see page 43).

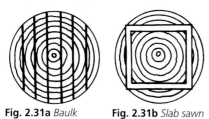

Fig. 2.31a *Baulk*   **Fig. 2.31b** *Slab sawn*

Logs are first quartered

Fig. 2.31c *Radial methods*

Fig. 2.30 *Tree felling in Canada*

Timber is a **hygroscopic** substance, taking in moisture from a damp atmosphere, but giving up moisture in a dry one. Consequently, damp wood shrinks in dry air and dry wood swells in damp air, which can affect the performance of manufactured products (for example, sticking doors). Unseasoned, **green timber** merely placed in a room for storage twists badly and joints open upon drying out. Seasoning aims to remove excess, unwanted sap and moisture from the wood. **Moisture content** (MC) is the amount of moisture contained in the wood and is expressed as a percentage of its dry weight, i.e:

$$\%MC = \frac{\text{initial weight} - \text{dry weight}}{\text{dry weight}} \times 100$$

Reducing MC to less than 20% increases the strength and stability of the wood and makes it more immune from rot and decay. There are two basic methods of seasoning. **Natural air seasoning** is traditional and fairly cheap to operate, but depends upon weather conditions. The converted timber is stacked as illustrated in Figure 2.32, within an open- or louvre-sided shed. A sloping roof protects against direct sun and rain.

Fig. 2.32 *Timber stacked for seasoning*

Artificial kiln seasoning provides a quicker, more controlled and reliable method. It offers a rapid turnover and is therefore used to process hardwoods. The timber is stacked as for natural seasoning, but on trolleys, before being put in the kiln (Figure 2.33). Steam is introduced which soaks and penetrates the timber. After a time, pressure and humidity are reduced and the steam is drawn out by fans. Heat is gradually introduced and the temperature raised. Finally hot, dry air circulates until the moisture content is reduced to the required level. Sampling, using test pieces, is undertaken at intervals to determine and meet the exacting requirements. Using this method, precise moisture content is obtained in a matter of one to two weeks per 25mm plank thickness, aiming for below 18% for general outdoor use, falling to below 14% for indoor use and around 10% for centrally-heated homes.

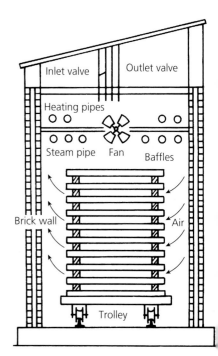

Fig. 2.33 *Cross-section of a drying kiln*

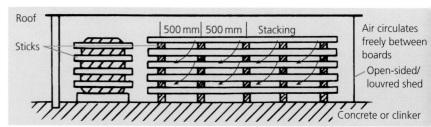

Defects in timber

Defects not only affect appearance, but also reduce the strength and durability of timber. The cause can be a variety of factors, including abnormal growth, wind damage, poor seasoning and attack by fungi and insects.

- **Shrinkage** affects the shape and movement of the board, depending upon how it was converted (Figure 2.34). Maximum shrinkage occurs along the direction of the annual rings (tangentially) with the medullary rays closing like a fan. There is negligible shrinkage in length.

- **Knots** are natural irregularities which format the junction of branches and may sometimes be considered defects. Healthy, **live knots** are only a defect if they are large in size or present in sufficient numbers. Loose, **dead knots** are an obvious source of weakness. **Irregular grain**, which can be spiral or twisted, is caused by distorted growth.

- **Interlocking grain** generally causes difficulties, such as tearing, in working.

- **Splits and shakes** (Figure 2.35) may be formed in logs. Radial splits follow the line of the medullary rays and may be formed before conversion. Similar splits in converted timber are called 'checks'. Shakes are separations in adjoining layers of wood. Thin hair-line cracks, which form across the grain in some timbers (such as mahogany) are called **thunder shakes**.

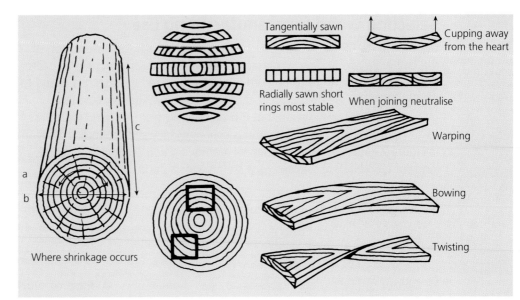

Fig. 2.34 *Movement and shrinkage*

Fig. 2.35 *Splits and shakes*

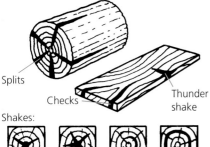

Splits

Checks

Thunder shake

Shakes:

Heart   Star   Cut   Ring

- **Fungal attacks** cause wood to decay. Oxygen and low temperatures are the necessary conditions which allow such attacks to develop. Fungi are parasitic plants whose spores cause the wood cells to collapse. Attacks result in loss of strength and weight and an increase in water absorption of the wood. **Dry rot** (Figure 2.36) is caused by the sponge-like fungus, Merulius lacrymons. It thrives in damp, unventilated situations where there is a lack of circulated air. Treatment involves complete removal and burning, and replacing with chemically treated timber.

Fig. 2.36 *Dry rot*

Wet rot affects external woodwork, for example window ledges and posts at ground level. The wood, when subjected to alternate wetness and dryness begins to decompose. Affected timber can be cut back to the remaining sound wood and replaced.

- **Stains** can be caused by otherwise harmless fungi, affecting mainly sapwood and lighter coloured hardwoods (e.g. sycamore).

- **Insect attacks** of wood in Britain are mainly restricted to **pin-hole boring insects**, which only attack timber in its green state, leaving small holes across the grain surrounded by dark stains. The four varieties of **beetle** illustrated in Figure 2.37 are common attackers of wood. Treatment to kill the beetles includes using proprietary chemicals, and in severe cases, fumigation by experts.

Fig. 2.37 *Insects that attack wood*

Common furniture beetle	Death watch beetle
Most common beetle, makes a honeycomb of tunnels in wood	Attacks oak and structural timbers in churches and historic buildings
Powder post (lyctus) beetle	House horn beetle
Damages the sapwood of hardwoods such as oak, ash and elm	Devours softwood, especially timbers in roof structures

Selecting wood for manufacturing use

Once converted, wood is identified by its characteristics, by observation of the end grain and microscopic examination of its cellular structure. In order to select the right wood to be used in a manufactured product it is useful to be able to recognise the different timbers available. Experience can be gained by careful observation, but more especially from working with this material. Consideration of the various characteristics and properties, shown in Figure 2.41, is useful as a means of identifying and classifying different kinds of wood before selecting one suitable for a particular item.

Fig. 2.38 *Cross-section of softwood (left) and hardwood (right)*

- **Weight** varies considerably, but generally hardwoods tend to be heavier than softwoods. Exceptions include yew, a heavy softwood and balsa, a very lightweight hardwood.

- **Colour** can be misleading, as wood does change when exposed to light. While many fade, others, such as teak, darken. It is probable that most softwoods are lighter in colour, but there are very many exceptions which include sycamore, holly and willow.

- The **smell** of wood fades with time, but can be revived by heat, including the friction created by planing. Many timbers have a distinctive smell, especially pine.

- **Grain**, including silver, figure, striped and ripple, adds to the unique nature of this material. Straight or irregular patterns are also keys to identification and attraction. Hardwoods being the more decorative, are preferred for furniture.

- **Texture** is another important consideration. Hardwoods offer more choice because more varieties are available. Their denser, closer grain formation provides a better surface where contact with food is concerned.

- For **durability**, hardwoods generally are more resistant to surface marking and have a longer life span.

- For **outdoor** use, selection is linked with rot-resistance as well as durability. Most softwoods deteriorate rapidly outdoors, but Western red cedar is a notable exception.

- **Ease of working** should be considered. Hardwoods are stated to be more difficult to work because they quickly blunt cutting edges. However, no certain rules apply as each piece of timber offers unique qualities. Yew, for example, is physically very hard, and knotty pine will certainly present problems when worked.

- The **costs** of using softwoods are considerably less than using hardwoods, with commercial investment in specific softwood varieties. Hardwoods are expensive, because of the time they take to mature.

- **Commercial forms and sizes** are available. After conversion and seasoning, timber is further reduced into smaller sections of standard shapes and sizes. It is sold either **rough-sawn** (also called nominal, full-size) or **ready-machined** (planer thicknessed). Planing can be **PBS** (planed both sides) or **PAR** (planed all round). Planed timber is described as the nominal (rough-sawn) size, but will actually be approximately 3mm smaller. Buying timber, especially selecting hardwood boards, can be an enjoyable experience, but remember about planed thicknesses and the need to make extra allowance for saw cuts and split ends!

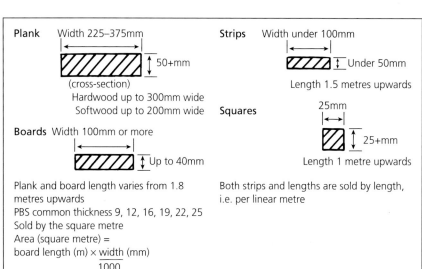

Fig. 2.39 *Standard preferred sizes in timber*

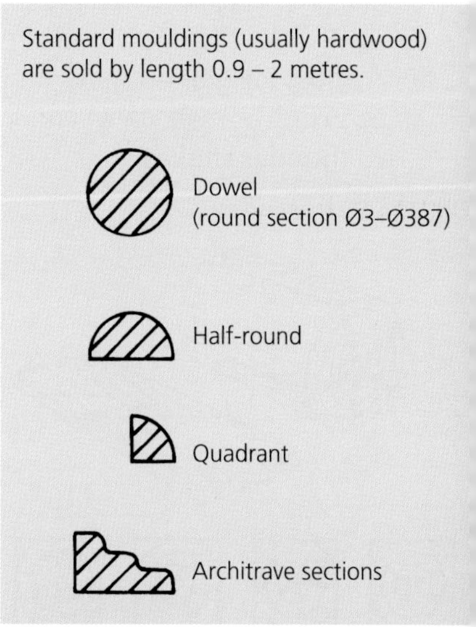

Standard mouldings (usually hardwood) are sold by length 0.9 – 2 metres.

Fig. 2.40 *Standard mouldings*

a) Hardwoods

NAME	ORIGIN/COLOUR	PROPERTIES AND WORKING CHARACTERISTICS	USES
Beech	Europe White to pinkish brown	Close-grained, hard, tough and strong, works and finishes well but prone to warping	Functional furniture (e.g. chairs, toys, tools, veneer, turned work, steam bending)
Elm	Europe Light reddish brown	Tough, durable, cross-grained which makes it difficult to work, does not split easily, has a tendency to warp, good in water	Garden furniture (when treated), turnery and furniture
Oak European English Japanese	Europe Light brown	Very strong, heavy, durable, hard and tough, finishes well, open-grained, it contains tannic acid which corrodes iron/steel fittings leaving dark blue staining in the wood, expensive	High-class furniture, fittings, boat building, garden furniture, posts, veneer
	Japan Yellow brown	Slightly milder, easier to work but less durable	Interior woodwork and furniture
Ash	Europe Pale cream colour and light brown	Open-grained, tough and flexible, good elastic qualities, works and finishes well	Tool handles, sports equipment, traditional coach building, ladders, laminating
Mahogany African (e.g. Sapele, utile)	Central-South America, West Indies, West Africa Pink reddish brown	Easy to work, fairly strong, medium weight, durable, available in long, wide boards, some difficult interlocking grain, prone to warping	Indoor furniture and shop fittings, panelling, veneers
Meranti	S.E. Asia Dark red, also white-yellow	Fairly strong, durable and fairly hard to work	(Mahogany substitute) interior joinery and furniture, plywood-red and white forms
Teak	Burma, India Golden brown	Hard, very strong and extremely durable, natural oils make it highly resistant to moisture, acids and alkalis, works easily but blunts tools quickly, darkens with exposure to light, very expensive	Quality furniture, outdoor furniture, boat building, laboratory equipment, turnery, veneers
Iroko	East/West Africa Yellow but darkens to dark brown	Like teak it is oily and durable, cross-grained, heavy	(Teak substitute) furniture cladding, construction work, veneers
Walnut African	Europe, USA, West Africa Yellow, brown, bronze, dark lines	Attractive, works well, durable, often cross-grained which makes planing and finishing difficult, available in large sizes	Furniture, gun stocks, furniture veneers
Obeche	West Africa Pale yellow	Straight, open-grained, soft, light and not very durable, sometimes cross-grained	Constructional uses, hidden parts of furniture, plywood core

b) Softwoods

NAME	ORIGIN/COLOUR	PROPERTIES AND WORKING CHARACTERISTICS	USES
Scots pine (red deal)	N Europe, Russia Cream, pale brown	Straight-grained, but knotty, fairly strong, easy to work, cheap and readily available	Mainly constructional work, joinery, paints well, needs outdoor protection
Western Red Cedar	Canada, USA Dark, reddish brown	Light in weight, knot free, soft, straight silky grain, natural oils make it durable against weather, insects and rot, easy to work, but weak and expensive	Outdoor uses, timber cladding of external buildings, also wall panelling
Parana Pine	South America Pale yellow with red/brown streaks	Hard, straight-grained, almost knot free, fairly strong and durable, smooth finish, tends to warp, expensive	Best quality interior joinery, i.e. staircases, built-in furniture
Spruce (whitewood)	N Europe, America Creamy-white	Fairly strong, small hard knots, resistant to splitting, some resin pockets, not durable	General indoor work, whitewood furniture (i.e. kitchens)

Fig. 2.41 Common timbers

MANUFACTURED BOARDS

Manufactured boards are valuable materials in their own right, with an important part to play alongside solid timber. Cost and property comparisons can be misleading, varying considerably according to type and grading. Many have no grain and must be considered separately from those that do. They can present their own problems of working, but advantages include their availability, in large, stable, standard sheets (1525 × 1525mm, 1220 × 2240mm), of uniform thickness and quality.

Veneer

The art of veneering (applying a thin sheet of decorative wood on to a plainer ground or base) can be traced back to ancient times. Development of this process took place in the seventeenth century, alongside the importation of exotic timbers. It forms the basis of the mass produced furniture of today, as well as the production of manufactured boards. It is an economic means of using exotic woods, many in short supply. Decorative effects can be created by **'matching' veneers** and making patterns by **'quartering'**, **cross-banding** and **marquetry** (Figure 2.42).

Fig. 2.42 *Matching veneers*

The first veneers were **saw cut**, thick and of high quality, but expensive due to the waste. They are still used for difficult and highly figured hardwoods, such as ebony. The majority of veneer is **knife cut**, using one of two methods. In each case

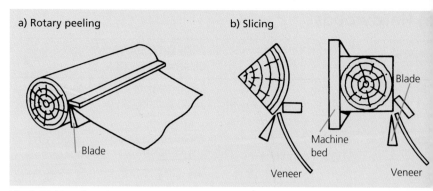

Fig. 2.43 *Manufacturing veneer*

the log is first prepared by steaming in water to condition (soften) the wood and give a clean cut. **Rotary peeling** involves mounting the prepared log on a machine similar to a lathe. It is slowly rotated and once it is cylindrical, a long knife is fed automatically into the log (Figure 2.43a) and a thin continuous sheet peeled off. It is then trimmed, cut dried and graded. This method produces the plainest veneer, 90% of which is used for manufactured boards. It is inexpensive with little waste. Decorative face veneers are produced by **slicing**. Prepared fitches (quarter, flat or half round) are secured on a movable frame, which is brought against a knife. Automatic resetting gives accurate, successive veneers, with closely matching grain (Figure 2.43b). **Constructional veneer** is usually thicker (1.5–2.3mm) and used for laminating; like all veneer it is sold by the square metre.

Plywood

The introduction of rotary cut veneers in the 1890s led to the exploitation of this material, and early uses included tea chests and piano frames. Plywood is formed using an odd number of thin layers (veneers) called **laminates**, with the grain of each running at right angles to its neighbour (Figure 2.44). This interlocking sandwich gives high uniform strength and resistance to splitting. The odd number of layers balance the stresses around the central core, cancelling out shrinkage across the grain and ensure the outer layers run in the same direction. In manufacture, rollers are used to apply adhesive and

then large presses and controlled heat cures (sets) the adhesive. Waterproof or resistant glue is used, making the bond stronger than the wood itself. The faces are cleaned, then graded by appearance quality, with defects repaired by circular plugs. Letters A, B and BB denote quality, for example B/BB would have one face better than the other, making it less expensive than B/B. Timber varieties include birch, alder, beech and gaboon. Standard plywood is commonly available white- or red-faced. Special forms include **veneered plywood** produced with a decorative face such as oak, mahogany, afromosia and teak. The reverse face usually has a compensating, less expensive veneer. **Marine plywood**, made from makore veneers and phenolic resin adhesive offer special protection for outside use including boat building. **Pre-formed plywood** is also available where rigidity and curved forms are needed. Other forms include weather-and-boil-proof (**WBP**) which is highly resistant to weathering, boil-resistant (**BR**) which is not suitable for prolonged exposure to weather, moisture-resistant (**MR**) and interior use only (**INT**).

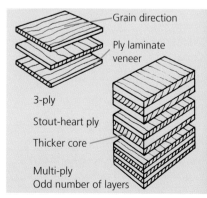

Fig. 2.44 *Common forms of plywood*

Blockboard

Blockboard is a collective term for a group of boards of similar construction, which although not as strong, possess many of the qualities of plywood. They provide cheaper substitutes when built up boards of greater thicknesses (12–25mm) are required.

Laminboard consists of a softwood core made up of parallel strips, 5–7mm in width, sandwiched between two outer veneers with their grain running at right angles to the core (Figure 2.45). The core is usually of pine or spruce, but in some countries hardwood is used. The facing veneer can be of birch or gaboon. Blockboard is similar, with wider core strips up to 25mm. The quality of both core and veneer can vary considerably. More expensive kinds are available, such as **superior 5 ply** which has additional veneers ensuring grain direction runs parallel to the core.

Chipboard

Chipboard is made of wood particles (flakes, chips and shavings) from all commercial timbers, bonded with synthetic phenolic resin. The small particles give no grain direction. The boards are produced by highly automated processes – either flat pressed under heat and pressure, or extruded between parallel metal patterns to produce core boards, which are less strong. The various market forms are illustrated in Figure 2.46. The **single-layer** type has

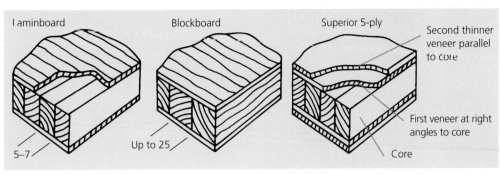

Fig. 2.45 *Types of blockboard*

interlocking similar-sized particles which give uniform appearance on all surfaces. **Sandwich construction** has coarser particles sandwiched between finer chips on the surface layers, giving a smoother surface finish. **Veneered chipboard** (**conti-board**) is more expensive and has both sides covered with hardwood veneer or plastic laminates.

Hardboards

Hardboards provide a cheap substitute for plywood where space filling, rather than strength, is required. Wood fibre, obtained from chips and pulped wood waste, provides the raw material. During manufacture it is exploded under high pressure, heat and steam to leave a fine, fluffy mass of brown fibres. These are refined and formed into mats that are thick felted blankets of loose fibres held together by natural lignum and other bonding agents. It is then pressed between steam-heated plates to give flat grainless sheets with one smooth, glossy face and a rougher textured surface. Conditioning follows, which involves adding moisture in a humidification chamber, to help

prevent warping. Hardboard has no regular grain and absorbs moisture very easily, making it unsuitable for outdoor work. **Tempered hardboard** is impregnated with oil and other chemicals during the pre-finishing process, making it stiffer, and giving it a harder finish which is more resistant to cuts, scratches and moisture. **Special hardboards** include types which are perforated, embossed, veneered and plastic-faced (Figure 2.47).

Medium density fibre board (MDF)

MDF is cost-effective and popular. It is generally more dense, thicker and heavier than hardboard, with smooth faces. Very stable and unaffected by humidity, it is a good electrical insulator and takes paint and other finishes very well.

Insulation board

Insulation board is, again, similar to hardboard, but not as compressed, has a low density and is both light and weak. It is a good insulator from heat and sound, and is suitable for interior walls. It is used for such items as acoustic tiles and notice boards.

All manufactured boards require some form of edge treatment, as shown in Figure 2.48.

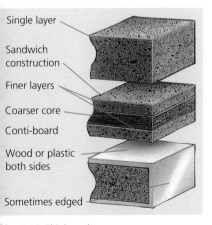

Fig. 2.46 *Chipboard*

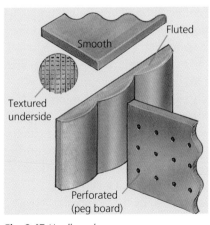

Fig. 2.47 *Hardboards*

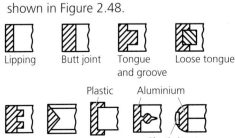

Fig. 2.48 *Edge treatment of manufactured boards*

FINISHES FOR WOOD

Basic preparation (cleaning-up) is necessary before any type of finish is applied to wood. A **smoothing plane** is used to give a clean smooth surface that is free from blemishes. Difficult hardwoods and thin veneers may require a **scraper** (Figure 2.49) using the burr to take off fine shavings.

Minor faults are removed with **glass paper**. This abrasive paper is made from different grades of ground glass glued to a paper backing. An alternative is **garnet paper**, made from a longer lasting, natural grit. Progressively finer grades are used, wrapped round a cork block, to ensure a flat surface. Always work in the direction of the grain, to prevent scratching. For large surfaces an electric orbital sander or industrial belt sander (with alumina oxide grits) can be used. Use a clean cloth to wipe away dust before applying any finish.

Finishes serve to prevent wood absorbing moisture, protect against decay and enhance appearance. Finishing is often carried out after final assembly but sometimes (e.g. finishing inside surfaces) it is best carried out prior to this. Use masking tape to cover joints and areas to be glued that do not require the application of a finish.

Staining

Staining (colouring) is used to enhance the grain and ensure an even colour match, with pigments for oak, walnut, etc. They can be water-based, or spirit-based aniline dyes, which dry more quickly, but oil-based stains are more versatile and longer lasting.

Oil

Oil brings out the natural appearance of wood, highlighting grain and colour. Regular application and subsequent coats congeal in the cells, building up a matt, water-resistant finish. Types include odourless, non-toxic **olive (vegetable) oil**, which is

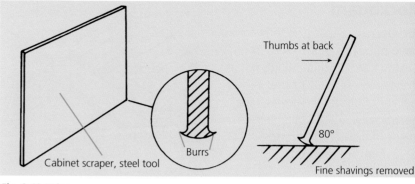

Fig. 2.49 *Using a scraper*

suitable for products such as bread boards or salad servers where there will be contact with food. **Teak oil**, made from linseed oil, golden size (a drying agent) and turpentine, is ideal for oily woods like teak, afromosia and iroko. Proprietary finishes (e.g. Sadolin) penetrate the surface to give lasting protection for exterior as well as interior softwood. They offer a more attractive alternative to creosote, as an outdoor preservative for fences and sheds.

Wax polish

A traditional finish, wax oil enhances the grain and produces a dull gloss shine. Beeswax, dissolved in turpentine forms a paste, which is applied using a cloth; adding carnauba or silicone wax increases the durability of the wood. Surfaces must first be sealed, using shellac (a natural resinuous product of the lac insect) dissolved in methylated spirits, or cellulose dissolved in solution, to penetrate the surface.

French polish

Traditionally, French polish was the most important finish for interior woodwork before the introduction of modern synthetic finishes. Time, skill and experience is needed to apply and build up successive coats of shellac (dissolved in methylated spirits). A variety of 'colours' such as transparent, white, button and garnet can be applied using a rubber (cotton wool wrapped in linen) with linseed oil as a lubricant. Quality is in part determined by the number of layers that are applied. The final one, called

'spiriting out', is done with methylated spirits to produce a high-gloss surface.

Cellulose lacquers provide a substitute for French polishing and are usually sprayed on. Being brittle, they can craze and crack and have a shorter life.

Synthetic resins

Synthetic resins (plastic varnishes) give a much tougher surface, which is heat-, water- and spirit-proof and capable of withstanding hard knocks. Available in clear, translucent or coloured shades it gives a high-gloss, satin or matt finish. It is best applied in thin coats using a brush or sprayer, gently rubbing down between coats with steel wire wool.

Paint

Paint provides a protective colouring for both indoor and outside softwoods. Knots need to be sealed with **knotting** (shellac) to prevent resin oozing out. Sharp corners should be slightly radiused with glass paper. First seal with a **primer**, then apply **undercoat**(s), rubbing down between coats with fine glass paper, prior to the final coat. Several types and colours are available, which conform to **British standards**, ranging from non-shine to high-gloss. **Emulsion paints**, vinyl or acrylic resin, are water based and cover well, but are neither waterproof nor very durable. **Oil-based** paints, many of which are sold in gel or non-drip form, are more durable and waterproof. **Polyurethane** paints harden when exposed to air, to give a tough scratch-resistant finish suitable for toys and furniture.

METAL

Metals make up a major portion of all the naturally occuring elements, and form about a quarter of the earth's crust by weight. All metals, with the exception of gold, are found chemically combined with other elements in the form of oxides and sulphates. **Ore** deposits are not distributed evenly, but tend to be concentrated in localised areas. Exploitation by open cast working (deep mining) depends on commercial considerations. The history of Cornish tin mining (Figure 2.50) illustrates how circumstances fluctuate and change.

Fig. 2.50 *A Cornish tin mine*

Ferrous metals

Ferrous metals have played an important part historically in human development and remain vital to our everyday life. They contain mainly ferrite (iron) with small additions. Almost all, including mild steel, cast iron and tool steel, are magnetic. **Iron** is the basis for the more sophisticated metals now available in this category. Britain has small deposits of **haematite ore**, but high-grade **magnetite ore** has to be imported. The conversion of ore into usable material involves a number of stages and processes: washing, grading and crushing. In the production of iron, ore is refined in a **blast furnace** to provide **pig-iron**.

Steel production involves reducing the carbon content of pig-iron from over 3%, to less than 1.5%, and in the case of mild steel to below 0.25%. The **basic oxygen furnace** is the principal method used for making large tonnages. The furnace (Figure 2.51) is tilted and charged with molten pig-iron and up to 30% scrap metal, then returned to a vertical position. A water-cooled lance blows oxygen into the melt at high speed, to combine with the carbon. Lime is also added as a flux, to help the impurities form a slag on the surface. Once the steel has been checked it can be tapped into ladles or poured directly into a continuous casting machine to produce billets, blooms and slabs.

The **electric arc furnace** (Figure 2.52) is used to produce high-quality (high-alloy) steels. The process involves the charge (selected pig-iron and scrap) being placed in a shallow refractory lined bath, with carbon electrodes lowered through the lid. The metal is melted by powerful arcs struck between the electrodes and metal, which produces high temperatures (over 3000°C). Impurities in the slag are removed by using lime and other deoxidisers, before pouring off by tilting. Alloying elements are added and test samples analysed before tapping into ladles.

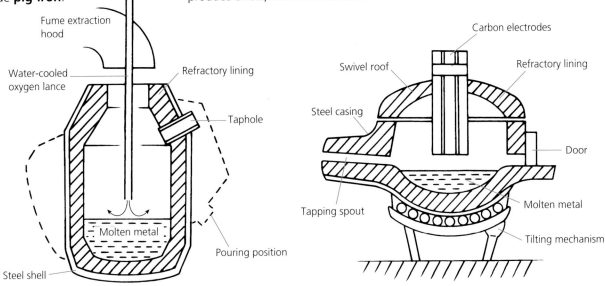

Fig. 2.51 *Basic oxygen furnace*

Fig. 2.52 *Electric arc furnace*

Non-ferrous metals

This group consists of metals such as copper, aluminium, tin and lead that contain no iron.

Aluminium is the most plentiful metal in the Earth's crust. Increasing demand for lightness combined with strength also makes it the largest, in terms of production output, within this category. **Bauxite**, a hydrated form of aluminium, is the only commercial source and is found in the USA, France, Australia and Africa. Unfortunately it is difficult to decompose and no cheap chemical is available, so an **electrolytic process** is needed for production, which reduces alumina to oxygen and aluminium. This is an expensive process as large quantities of electrical energy are consumed.

The **reduction cell** used in the process (Figure 2.54) is a shallow steel box with a carbon lining, which acts as a negative electrode. Suspended above it are a number of carbon rods which act as anodes. Mixing with **molten cryolite** reduces the melting point and dissolves the alumina. A powerful electric current, passed through the mixture, causes the aluminium to be liberated. It sinks and is deposited on the carbon lining. Periodically, the very pure (99%) aluminium is syphoned off and cast into ingots ready for further processing.

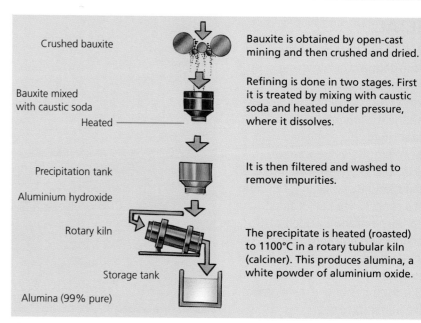

Crushed bauxite — Bauxite is obtained by open-cast mining and then crushed and dried.

Bauxite mixed with caustic soda
Heated — Refining is done in two stages. First it is treated by mixing with caustic soda and heated under pressure, where it dissolves.

Precipitation tank — It is then filtered and washed to remove impurities.

Aluminium hydroxide

Rotary kiln

Storage tank — The precipitate is heated (roasted) to 1100°C in a rotary tubular kiln (calciner). This produces alumina, a white powder of aluminium oxide.

Alumina (99% pure)

Fig. 2.53 *The production of aluminium: Stage 1*

To improve the properties of hardness and strength, a large proportion of the pure metal is alloyed with other metals such as copper, manganese and magnesium. As much as five times the amount of electrical energy is needed to produce 1 tonne of aluminium, compared to that required to produce 1 tonne of steel. Like steel, it is available in a variety of forms (see Figure 2.55).

Alloys

Alloys are mixtures of two or more metals formed together with other elements to create new metals with improved properties and characteristics. There are two groups: ferrous alloys and non-ferrous alloys.

Examples of ferrous alloys are **stainless steel** (steel and chromium) and **high-speed steel** (steel and tungsten). **Non-ferrous alloys** include brass (copper and zinc) and duralumin (aluminium and copper).

Properties of alloy steel, such as hardness, are increased by the addition of other metals such as chromium, tungsten, nickel and vanadium. Unlike carbon steels, which lose hardness at high temperatures, high-speed steels maintain hardness even at red heat and are used to make cutting tools and drills. Stainless steel contains 12% chromium and some nickel and offers qualities such as resistance to corrosion.

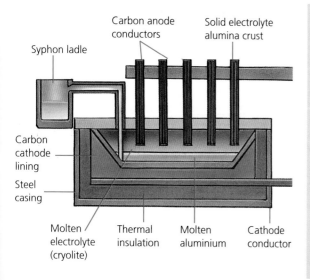

Carbon anode conductors   Solid electrolyte alumina crust

Syphon ladle

Carbon cathode lining

Steel casing

Molten electrolyte (cryolite)   Thermal insulation   Molten aluminium   Cathode conductor

Fig. 2.54 *The electrolytic reduction cell*

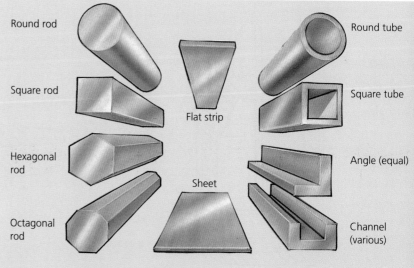

Round rod   Round tube

Square rod   Square tube

Flat strip

Hexagonal rod   Angle (equal)

Sheet

Octagonal rod   Channel (various)

Fig. 2.55 *Commonly available metal forms*

METALLIC STRUCTURE

All **atoms** have a similar basic structure, but they differ in size and weight. Orbiting around the nucleus (centre) of the atom are **electrons**, which are light particles with a negative electrical charge (Figure 2.56). Atoms determine how elements behave, both physically and chemically. An atom or group of atoms that has lost or gained one or more electrons to become electrically charged are called ions. Atoms and ions are attracted to one another by electrical forces. This attraction between atoms or ions, which holds them together in a molecule (a crystal made up of at least two atoms) is called **metallic bonding**. Metals usually have only one or two electrons attached very loosely in their outer electron shell. It is the free movement of electrons which give metals their high thermal and electrical conductivity. This mobility also gives rise to plasticity (malleability and ductility). **Ionic bonds** form the building blocks of some crystal structures; as soon as a bond between ions is broken, another is formed. All metals, with the exception of mercury, are solid at normal atmospheric temperature. In their molten (liquid) form, metal atoms are highly energised and move around at random. When atoms solidify (freeze), a regular structure is adopted. They group and pack closely together in geometric patterns (**crystal lattice**

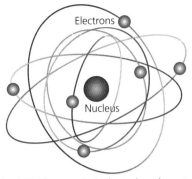

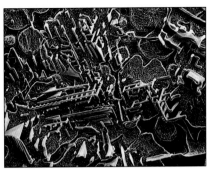

Fig. 2.56 *The atom is similar to the solar system*

structures). Most metals crystallise into one of the three basic types of lattice illustrated in Figure 2.57.

Iron is unusual because it changes from a body-centred cubic structure (**BCC**) to a face-centred cubic structure (**FCC**) when heated to 910°C. Taken beyond 1400°C it reverts back to BCC once again. In cooling, the changes take place in reverse. In its FCC form (Figure 2.58) carbon is absorbed, which important in steel making.

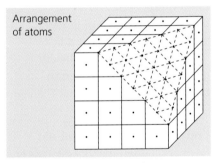

Fig. 2.58 *FCC Face-centred cubic structure*

Pure metals solidify at a fixed temperature, beginning with the formation of minute **seed crystals** (nuclei). The cell units grow in all directions, forming small skeletal crystals called **dendrites** (Figure 2.59).

Fig. 2.59 *Dendrites in a sample of high-speed steel*

They develop independently until contact is made with neighbouring growths. Finally, small crystals (**grains**) are formed as the atoms between the arms of individual dendrites become attached to the structures. Dendrite growth always causes irregularities at the grain boundaries due to the different orientation of the original nuclei.

Finally only grain boundaries are visible

Fig. 2.60 *Stages in the solidification of metal*

The crystalline nature of zinc, often used to 'galvanise' (coat) steel, is clearly visible in Figure 2.61.

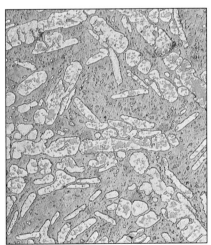

Fig. 2.61 *Crystalline structure of zinc*

Some crystals, however, are so small that they need to be examined under a microscope. A metal's physical properties depend not only upon crystal and grain structures, but also on the presence of defects within the grains. Flaws or faults, caused by interruptions in the pattern, are called **dislocations**. Atoms can be completely missing or misaligned.

CPH Close packed hexagonal structure	FCC Face-centred cubic structure	BCC Body-centred cubic structure
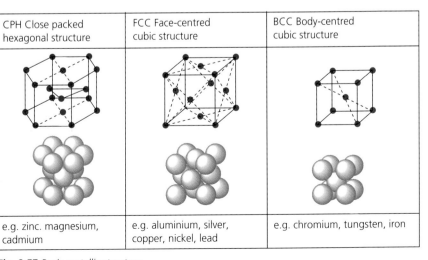		
e.g. zinc. magnesium, cadmium	e.g. aluminium, silver, copper, nickel, lead	e.g. chromium, tungsten, iron

Fig. 2.57 *Basic metallic structure*

HEAT TREATMENT

The process of heating and cooling metal can be used to alter its properties and characteristics. Those occuring at the grain boundaries are different from those within the grains. To obtain maximum strength for work at low temperatures, fine grains with many boundaries are needed. Working at higher temperatures the reverse is true, that is large grains with fewer boundaries are needed. The size of grains, which affects strength, is also controlled by the rate of cooling. **Cold working** (hammering, bending and rolling) introduces changes by applying stress, which results in **work hardening**. Before further work can take place, some restoration of the crystal structure is required.

Annealing

Annealing is the process whereby heat is introduced to mobilise the atoms and relieve internal stresses. This takes place at relatively low temperatures, but as the temperature increases, a point is reached where new crystals begin to grow. Grain growth and ultimate grain size depend on the length of time and the temperature of the treatment. Most annealing involves complete recrystallisation of the distorted structure (Figure 2.62).

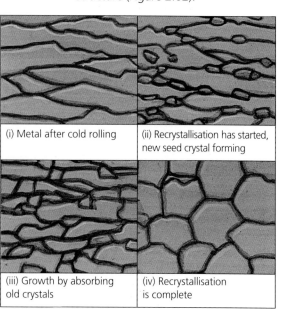

(i) Metal after cold rolling

(ii) Recrystallisation has started, new seed crystal forming

(iii) Growth by absorbing old crystals

(iv) Recrystallisation is complete

Fig. 2.62 *Recrystallisation taking place during the annealing process*

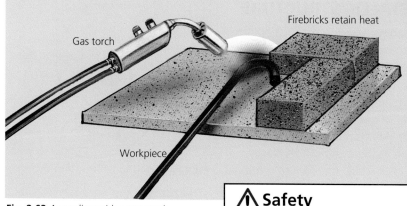

Fig. 2.63 *Annealing with a gas torch*

Ferrous metals, for example bright steels, are heated to **bright cherry red** (around 725°C), **soaked** (retaining the temperature for a period of time) and then allowed to cool very slowly. The result is the formation of large coarse grains, which give the metal a soft, workable quality.

With **non-ferrous metal**, such as **aluminium**, there is a danger of overheating because of its low melting point (660°C). It is first rubbed with soap before being **gently heated** until the **soap blackens** (350–400°C), and then left to cool.

Copper is heated to a **dull red heat** and either quenched in water, or allowed to cool in air.

Brass is heated to a **dull red heat** and allowed to cool slowly. Both copper and brass form scales when heated which can be cleaned off **chemically** by 'pickling' (placing the cooled metal in a bath of dilute sulphuric acid).

⚠ Safety

Pickling needs to be carried out under carefully controlled conditions. Extreme care should be taken to avoid contact with the acid. Remove carefully with tongs (brass) and wash carefully in water.

An alternative **mechanical method** of removing the scale involves rubbing with pumice powder and steel wool.

Normalising

This process is confined to steel. It is used to refine grains, which have become coarse through work-hardening, to improve the ductility and toughness of the steel. It involves heating the steel to just above its **upper critical point** (see Figure 2.64). It is soaked for a short period then allowed to cool in air. Small grains are formed which give a much harder and tougher metal with normal tensile strength and not the maximum softness achieved by annealing.

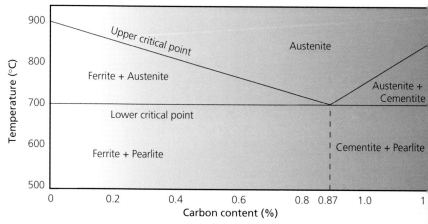

Fig. 2.64 *Heat treatment of steel*

Hardening

The physical properties of steel vary considerably according to the amount of carbon present with the iron. There are a number of forms.

Ferrite is a solid solution (BCC) iron which is soft and ductile.

Cementite is a chemical combination called iron carbide which is hard and brittle.

Pearlite, made up of alternate layers of ferrite and cementite, with 0.87% carbon, is very strong.

Austenite, a solution of carbon in (FCC) iron with a maximum 1.7% carbon content is soft, and it exists only at about 700°C.

Steel made up of pearlite provides maximum strength with hardness and is used to make cutting tools. However, it must first undergo changes. Maximum hardness, together with small grain size, is achieved by heating uniformly to just over 720°C for steels with 0.87% or more carbon, and to just above the critical point for steels with less than that amount (see Figure 2.64). It is soaked at this temperature long enough for the whole mass of metal to change into austenite. If it is immediately quenched in water, it becomes very hard and brittle. The iron changes back to BCC (from FCC in austenite) before the carbon has had time to escape. In this frozen state the metal is called **martensite**. To minimise distortion and prevent cracking it is quenched vertically in cold water (or salt water for heavy

sections and plain carbon steel, or in oil [with high flash point] for most alloy steels). If quenching is delayed, complete transformation to martensite does not take place.

The degree of hardness depends upon the amount of carbon present in the steel and the form in which it is trapped on quenching. The full effects are only possible in steels with a carbon content above 0.8%, known as high carbon or plain carbon steels. Mild steel, below 0.4% carbon content, cannot be hardened in this way. Between the two are medium carbon steels which gain a degree of toughness, rather than hardness. After hardening, the metal is resistant to wear, but it is also brittle and easily broken under load. By reducing the hardness slightly, a more elastic, tougher material, capable of retaining a cutting edge, is produced. This process is called tempering.

Tempering

In the workshop tempering involves cleaning the hardened steel to brightness with emery cloth, so oxide colours will be visible. The material is heated to 230–300°C, depending on use. For example, a lathe tool subject to steady pressure can be left harder than a chisel which will receive intermittent blows, or a screwdriver which has to withstand torque stresses (see Figure 2.65).

When bright steel is heated, well behind the cutting edge, by a gas air torch (see Figure 2.66) coloured

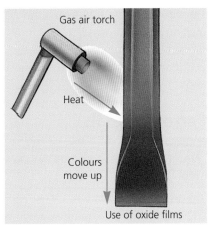

Fig. 2.66 *Tempering a cold chisel*

oxides develop. As the desired colour reaches the tip, it is immediately quenched in cold water. In industry, temperature-controlled ovens are used, which makes quenching unnecessary.

Case hardening

Case hardening is a process used with mild steels to give a hard skin. Heated to **cherry red**, the metal is dipped in **carbon powder** (Figure 2.67a) then reheated and dipped two or three times more before finally reheated and quenched in water to harden the skin. This is used where only the surface is subject to wear and a soft core is needed to withstand sudden shocks in equipment such as tool holders. Similarly **carburising** (Figure 2.67b) involves placing the mild steel specimen in a box packed with charcoal granules, heating in an oven to 950°C and allowing to soak for several hours. This thickens the outer skin, making it hard while the inner core remains soft.

Colour	Hardest	Approx temp (°C)	Uses
Pale straw	Hardest	230	Lathe tools, scrapers, scribers
Straw		240	Drills, milling cutters
Dark straw		250	Taps and dies, punches, reamers
Brown		260	Plane irons, shears, lathe centres
Brown-purple		270	Scissors, press tools, knives
Purple		280	Cold chisels, axes, saws
Dark purple		290	Screwdrivers, chuck keys
Blue	Toughest	300	Springs, spanners, needles

Fig. 2.65 *Guidelines for tempering*

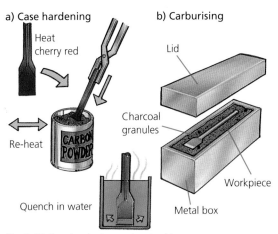

Fig. 2.67 *Case hardening and carburising*

The table in Figure 2.68 shows the melting point, composition, properties and working characteristics and the uses of ferrous, non-ferrous and alloy metals.

KEY

Suitability for project work

 poor   average   ◆ ◆ ◆ good

Ferrous metals

NAME AND MELTING POINT	COMPOSITION	PROPERTIES AND WORKING CHARACTERISTICS	USES
Cast iron 1200°C	Iron + 3.5% carbon, wide range of alloys, white, grey and malleable forms	Hard skin, brittle soft core, strong under compression, self lubrication, cannot be bent or forged ◆	Heavy crushing machinery Car brake drums or discs Vices or machine parts
Steel Mild steel 1600°C	Alloys of iron and carbon 0.15–0.35% carbon	Tough, ductile and malleable, high tensile strength, easily joined, welded, poor resistance to corrosion, cannot be hardened and tempered, general purpose material ◆ ◆ ◆	Nails, screws, nuts and bolts, Girders Car bodies
Medium carbon steel	0.4–0.7% carbon	Strong and hard, but less ductile, tough or malleable ◆ ◆	Garden tools (trowel, fork) Springs Rails
High-carbon steel (silversteel) 1800°C	0.8–1.5%	Very hard, but less ductile, tough or malleable, difficult to cut, easily joined by carbon heat treatment, strength decreases above 0.9% ◆ ◆	Hand tools (hammers, chisels, screwdrivers, punches)
Alloy steels Stainless steel	Alloys 18% chromium 8% nickel 8% magnesium	Hard and tough, resists wear, corrosion-resistant, different forms affect malleability (types 18/8), difficult to cut or file ◆	Sinks Cutlery Dishes, teapots
High-speed steel	Medium carbon steel + tungsten + chromium + vanadium	Very hard, resistant to frictional heat even at red heat, it can only be ground ◆	Lathe cutting tools Drills Milling cutters
High tensile steel	Low carbon steel + nickel	Corrosion-resistant, low rate of expansion, exceptional strength and toughness ◆	Gears/engine valves Turbine blades
Manganese steel	1.5% manganese	Extreme toughness ◆	Chains Hooks and couplings

Fig. 2.68 *Metals*

b) Non-ferrous metals and their alloys

NAME AND MELTING POINT	COMPOSITION	PROPERTIES AND WORKING CHARACTERISTICS	USES
Aluminium 660°C	Pure metal	High strength/weight ratio, light, soft and ductile (FCC), work hardens in cold state, annealing necessary, difficult to join, non-toxic, good conductor of heat and electricity, corrosion-resistant, polishes well ◆ ◆ ◆	Kitchen cooking utensils (pans) Packaging, cans, foils Window frames
Casting alloy (LM 4) (LM 6)	3% copper 5% silicon 12% silicon	Casts well, sand and die casting, good machineability, tougher and harder, increased fluidity ◆ ◆ ◆	Engine components, cylinder heads
Duralumin	4% copper 1% manganese + magnesium	Almost the strength of mild steel but only 30% of the weight, hardens with age, machines well after annealing ◆ ◆	Aircraft structure
Copper (Cu) 1083°C	Pure metal	Malleable, ductile (FCC), tough, suitable for hot and cold working, good conductor for heat and electricity, corrosion-resistant, easily joined, solders and brazes well, polishes well, rather expensive ◆ ◆ ◆	Hot water storage cylinders Central heating pipes/tubing Wire electrical Copper clad board (circuits)
Copper alloys Gilding metal	15% zinc	Stronger, golden colour, enamels, easily joined ◆ ◆ ◆	Architectural metalwork Jewellery
Brass 900–1000°C	35% zinc	Corrosion-resistant, increased hardness, casts well, work hardens, easily joined, good conductor of heat and electricity, polishes well ◆ ◆ ◆	Casting (e.g. valves) Boat fittings Ornaments
Bronze 900–1000°C	10% tin	Strong and tough, good wearing qualities, corrosion-resistant ◆ ◆	Statues Coins Bearings
Tin (Sn) 232°C	Pure metal	Soft and weak, ductile and malleable, excellent resistance to corrosion even when damp, low melting point ◆ ◆	Bearing metals Solder
Tin plate	Steel plate tin coated	Bends with mild steel core, non-toxic ◆ ◆	Tin cans
Lead (Pb) 327°C	Pure metal	Very heavy, soft, malleable and ductile but weak, corrosion-resistant, even by acid, low melting point, casts well, electrical properties ◆ ◆ ◆	Roof coverings – flashings Plumbing Insulation against radiation
Zinc (Zn) 419°C	Pure metal	Very weak, poor strength/weight ratio, extremely resistant to atmospheric corrosion, low melting point, ductile (CPH) but difficult to work, expensive ◆	Galvanised steel, dustbins Corrugated iron sheet roof Die casting alloys and rust proof paints

FINISHES FOR METAL

Finishes are used for both protection and decoration. Left unprotected, metals become dull, tarnish and if the conditions are right, they start to oxidise. It is essential that metal is carefully prepared by **cleaning-up** to remove all working marks that would otherwise spoil the finish. **Non-ferrous metals** can be chemically cleaned by **pickling** in an acid bath (comprising 1 part sulphuric acid to 10 parts water). This removes dirt, grease and any oxide film. After removal (with brass tongs) it is washed in water before further cleaning with a damp cloth, or steel wool, dipped in **pumice powder**. More persistent blemishes can be removed with a wetted **water-of-Ayr stone**.

Ferrous metal finishing

On steel a bright finish is achieved by **drawfiling** in one direction. **Emery cloth** (graded abrasive grit bonded on a cloth backing) is wrapped around the file to give a high standard of finish. Smearing with petroleum jelly or light grease provides some protection from rust.

Steel can be **blued** by heating the surface until a blue oxide forms and then quenching in water. After wiping dry, applying beeswax offers limited protection for decorative work. **Blackening** also gives good corrosion resistance to steel artefacts and is extensively used on forgework. It involves burning oil on to the surface of the steel by heating to dull red heat, then quenching in a high flash-point oil.

Fig. 2.70 *Forgework*

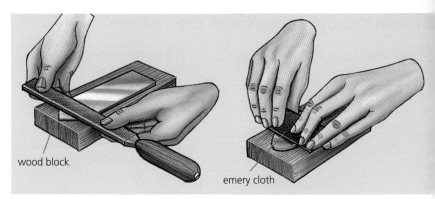

Fig. 2.69 *Drawfiling*

Painting

For painting metal, the surface must be thoroughly cleaned and **degreased** using paraffin, then washed with hot water and detergent. A zinc chromate primer is suitable for steel. For maximum protection an oil-based undercoat and topcoat should also be used. **Hammerite** offers a one-coat, quick drying protective solution for ferrous metals and is available in a range of colours with a crackle, hammered or smooth finish.

Plastic coating

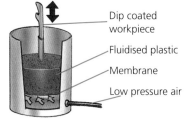

Dip coated workpiece
Fluidised plastic
Membrane
Low pressure air

Fig. 2.71 *The fluidisation tank*

Plastic coating is suitable for most metals and is used for coating metal baskets, crates, racks, tool handles and fittings. The metal needs to be thoroughly cleaned, degreased and heated evenly in an oven to about 180°C before the coating process (called **fluidisation**) takes place in a special tank (Figure 2.71). The work is then plunged quickly into the fluidised powder (using tongs or holding wire) and left for a few seconds while the powder sticks to the metal to give a thin coating. The work is then returned to the oven to completely fuse the coating and leave a smooth glossy finish. This process is used for coating baskets, crates, racks, tool handles, etc.

Polishing

Polishing may be done by hand, using metal polish, or by using a **buffing wheel**. Mops, made of soft stiff felt, calico (linen) or very soft swansdown for final finishing, are first dressed with a proprietary polishing compound. This consists of fine grits bedded into a wax bar but varies according to the material and finish required. The work is held firmly and pressed against the revolving lower half of the mop.

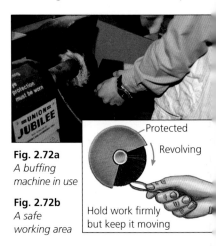

Fig. 2.72a *A buffing machine in use*

Fig. 2.72b *A safe working area*

Protected
Revolving
Hold work firmly but keep it moving

Lacquering

Lacquering provides a barrier against tarnishing and oxidising after polishing. A thin layer of cellulose, gum or varnish is brushed or sprayed on, to leave a transparent coating which allows the natural colour to show through. It may be necessary to use **white spirit** to degrease the surface first. This finish is particularly suitable for decorative items in brass and copper, including jewellery. Metals can also be **chemically coloured**, using potassium sulphide (brown) or ammonia and sodium acetate (green) prior to lacquering.

Anodising

Anodising is a finishing process associated with **aluminium** and is used to thicken the natural oxide film. The workpiece is used as the + (positive) anode, with lead plates forming the – (negative) cathode. All of this is immersed in a solution of sulphuric acid, sodium sulphate and water. A thin oxide skin forms when a direct current (DC) is passed through the solution. The film is converted by boiling in clean water, at which stage a **colouring dye** may be added, before the final surface is protected with lacquer. This type of finish is usually to be found on a range of aluminium items.

Fig. 2.73 *Anodised storage jars*

Enamelling

Enamelling uses powdered glass which is melted to flow over the metal to give a hard colourful, decorative and protective finish. Vitreous (stove) enamelling is used on steel for equipment such as cookers to provide a finish resistant to heat, chemicals, wear and corrosion. However, it can be chipped and damaged by physical knocks and mistreatment. A similar process is extensively used on copper, gilding metal and silver for decorative purposes, especially on jewellery.

Fig. 2.74 *Enamelled jewellery*

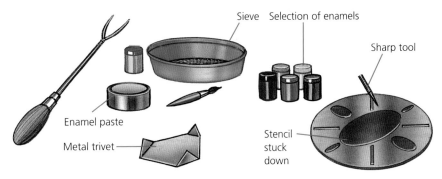

Fig. 2.75 *Enamelling equipment and techniques*

After degreasing, the surface is brushed with a thin adhesive (polycell paste) which burns away when heated. Selected powdered glass is sieved to cover the adhesive and the work is transferred and placed on a suitable trivet, or directly within a kiln. Heated to about 750°C, the enamel melts and quickly starts to flow, at which point it is taken out and left to cool.

Fig. 2.76 *An enamelling kiln*

Decorative techniques include the use of stencils, and **scrolling** using glass lumps and thin rods. More advanced techniques include **cloisonné** where thin wire is built up on the surface of the metal and the enclosed spaces filled with different coloured enamels.

Etching

Etching is a finishing process which enables designs and patterns to be made in the surface of the metal using an acid (the etchtant). Removing the resist, by scratching with a sharp instrument, exposes the metal directly to the etchant. An acid resist (e.g. paraffin wax) is applied to the parts that are not to be etched, or a stencil can be used to mask out those parts. **Nitric acid**, or for safety, a strong alkaline, **ferric chloride** is used

to eat away the copper. A feather can be used to brush away any debris which might restrict the chemical action. Alternatively, inverting with a piece of plasticine, as shown in Figure 2.77 also prevents this.

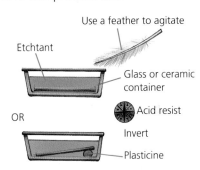

Fig. 2.77 *Etching techniques*

Contrast can be achieved between the matt/textured etched area and the shiny surface once the wax is removed and the surfaces lacquered.

Electroplating

Electroplating is used to give base metals, like copper, a coating of another protective or more decorative metal, such as silver. It is similar to anodising, except the work itself is the – cathode. The work is suspended in an **electrolyte**, a solution of the appropriate metallic salt, which acts as the conductor. A rod or plate of the metal to be deposited, provides the + anode.

Fig. 2.78 *Electro-plated chrome taps*

PLASTICS

This group of materials is not easy to define because it covers a wide range of diverse substances. A basic characteristic is that at some stage the material is putty-like ('**plastic**'): a state which is neither solid nor liquid, but somewhere between. At this stage shaping and moulding by heat and pressure takes place before **setting** into the desired form. Natural 'plastic' material occurs in the form of resins, clays, animal horns, amber, bitumen and shellac. Natural waxes and resins have been used for centuries as seals for documents, making use of an important property – its ability to be moulded.

The introduction of rubber in the 1820s and the later discovery of vulcanisation, which enabled the rubber to be moulded into shape, led to further developments in synthetic plastic materials. The first to be introduced, **Parkensine** (cellulose nitrate), a horn-like material, was made from natural cellulose and nitric acid. Another, **celluloid**, was developed from cellulose fibre of wood and cotton. By the end of the century there were only a few plastics commercially available and they were generally used to imitate natural materials such as horn, ivory, coral and tortoiseshell. Later **Bakelite**, a thermosetting plastic based on phenol and formaldehyde had much commercial success in conjunction with the developing electrical industry.

Fig. 2.80 *Typical use of modern plastics*

Fig. 2.79a *A celluloid doll's head from 1910*

The period leading up to the Second World War provided the stimulus needed for the mass production of synthetic plastics, the forerunner of the huge, thriving plastics industry of today. A synthetic rubber called **styrene** led to **polystyrene**, and **vinyl** provided a whole range of plastics including **PVC**. Clear **acrylic** was introduced through its use in aircraft canopies, with further research resulting in **polythene**. Meanwhile **nylon**, a strong resilient fibre capable of being spun when molten, offered a wide range of uses including clothing, propellers and bearings.

Its introduction initially as a substitute for more expensive raw materials, has unfortunately led the term 'plastic' to be associated with cheap, imitation commercial products. Today however, this is far from being the case. Many quality items could not be made from anything else. The unique range of properties this group of materials has to offer is now recognised, appreciated and extensively exploited. Plastic products are to be found in every home in kitchen equipment, floor coverings, electrical equipment and so on. More things than you may realise are made from plastics. Many have long complex chemical names such as polymethyl methacrylate, which is commonly known as acrylic. Hard and brittle, soft and flexible, there are hundreds of different types and even the same types of plastic can differ in texture and appearance. Polystyrene, for example, is produced as expandable blocks and sheets as well as high density moulded forms.

Fig. 2.79b *A bakelite telephone*

The potential and the scope of plastics affect every aspect of our lives, from healthcare to space exploration. There are two main sources of plastic (see Figure 2.81): **natural** sources whose modified forms play only a small part in the industry, and **synthetic** sources. The latter is the major supplier of the raw materials that go in to the production of plastics, mainly by refining crude oil. The fraction used for making plastics is the hydrocarbon **naphtha** (fractions are the parts of crude oil extracted during the refining process – tar and petrol are also fractions of crude oil). The liquid is heated with steam to break up its structure, **cracking** it into fragments, the most important being ethylene and propylene.

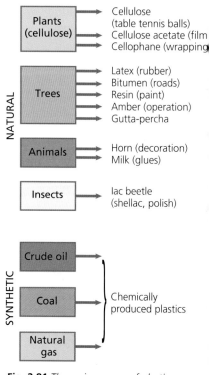

Fig. 2.81 *The main sources of plastics*

THE STRUCTURE OF PLASTICS

Molecules are the smallest units which form the building blocks of chemical compounds. In naturally occurring compounds they are short and compact, consisting of only a few atoms joined together.

In plastics, molecules do not stay as single units, but link up to form long chains of **giant molecules**. The small units forming the links (similar to a bicycle chain) are called **monomers**. The process of linking these units to form the chain is known as **polymerisation**. A **polymer** ('poly' meaning 'many'; 'mer' meaning 'unit') consists of between 200 and 2,000 units in the molecule. **Carbon** forms the backbone of polymer chemistry, but other elements, notably hydrogen, oxygen, nitrogen, fluorine and chlorine, also play a part. Compounds with one or more carbon atoms linked chemically, by double bonds, are called **unsaturated** compounds. They are ideally suited for linking together to form polymer chains in a process known as **additional polymerisation**. In **ethylene** (Figure 2.82) the carbon atoms link with two hydrogen atoms, but have spare electrons to link themselves together, **double-bonding**, to form long chains. As chains form they become entangled and bond together through weaker van der Waals forces, making solid material.

Plastics and polymers can also be formed by **condensation polymerisation**. This involves two different monomers reacting to produce a larger chain molecule (a **macromolecule**). Parts of the smaller molecules are split off as a by-product, usually water (H_2O). An example of this process is **Bakelite**, one of the first patented synthetic plastics, which is a thermosetting material made from phenol and formaldehyde (Figure 2.83). The process is also used to produce thermoplastics, notably **nylon** (polyamide).

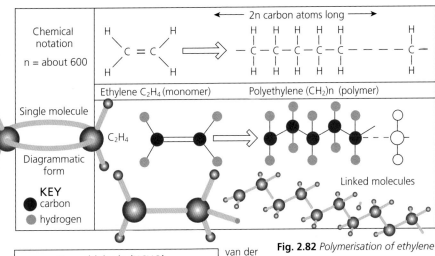

Fig. 2.82 *Polymerisation of ethylene*

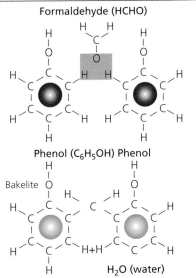

Fig. 2.83 *Condensation polymerisation*

The reaction to heat is a key determining factor common to the two main groups of plastic.

Thermoplastics

The thermoplastics group includes the majority of plastics in common use and range from the rigid to extremely flexible. Thermoplastics include polythene, polypropylene and polystyrene. They are made from long chain molecules. Entangled but flexible, rather like cooked spaghetti (Figure 2.84) they are held together by mutual attraction (van der Waals forces). These can be lessened by the introduction of heat energy. Distance between the molecules increases with movement, causing them to untangle and become soft, pliable and easily moulded. When the heat is removed, the chains cool, reposition and the material becomes stiff and solid once more. This **'plastic memory'** can be

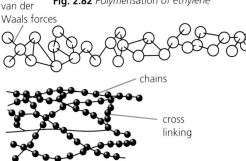

Fig. 2.84 *Entangled chains (thermoplastics – top) and covalent cross-linked chains (thermosets – bottom)*

repeated many times, as long as no damage is incurred by overheating.

Low- and high-density polythene is chemically the same. The differences are caused by the amount of branching within the chains. Branching hinders close packing, while linear molecules move more readily and become ordered and aligned. Branching also effects **crystallinity**, increasing stiffness and hardness in high-density polythene.

Thermosetting plastics

Thermosetting plastics also begin as long chain-like molecules, but they become chemically tied by **covalent bonds** (sharing of electrons) and are **cross-linked** when the polymer is heated, usually under pressure. They set with heat and have little plasticity. The molecules link side by side as well as end to end (Figure 2.84). Once the rigid network structure has formed it cannot be reheated and changed. This group is rigid and non-flexible even at high temperatures and includes polyester resin and urea formaldehyde.

PROPERTIES OF PLASTICS

Even a walk along the beach after high tide has left its debris reveals many important properties of this group of materials. Plastic containers contribute to environmental pollution because of their low density and resistance to corrosion and decay. Each group of plastics has particular characteristics. **Thermoplastics** are easily moulded and economic because waste can be re-used. However, where heat is applied they are less useful. Many soften and lose their rigidity at temperatures over 100°C (an exception being PTFE). By contrast, **thermosetting plastics** withstand higher temperatures without losing rigidity and are good insulators.

Mechanical properties

Strength is generally much lower in plastics than other structural materials, but lightness gives a good strength: weight ratio. It also varies considerably with temperature.

Corrosion resistance is a valuable consideration with these materials. Although not indestructible, many are able to withstand impact and misuse.

The specific properties of **nylon** and **PTFE**, having low **coefficients of friction**, make them ideal materials for bearings. The **disadvantages** are that they are prone to **deformation**, especially a tendency to **creep** (elongate under load). Deformation depends upon the temperature they are exposed to, but deterioration is rapid above 200°C. Many polymers become brittle and develop surface cracks when exposed to ultra-violet light.

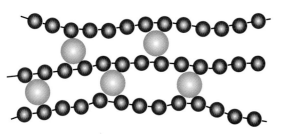

Fig. 2.86 *Plasticisers act as spacers*

Fig. 2.85 *Products made from a range of plastics*

Mechanical properties are improved by the use of **additives**. Liquid **plasticisers** improve flow properties, by providing mechanical spacers to separate the polymer chain and reduce the forces of attraction (Figure 2.86). Added to **PVC** they lower the softening temperature making it less brittle and, therefore, more suitable for packaging or wall coverings than drain pipes.

Catalysts are chemical peroxides used to increase or decrease the speed at which molecules link up. **Accelerators** (e.g. cobalt napthanate) are chemicals used to shorten the setting time of resin. **Fillers**, usually in the form of powdered solids which do not mix chemically, are used with thermosetting compounds. They reduce costs by providing low cost bulk (rather than an expensive polymer) and improve strength by increasing dimensional stability and impact resistance. **Colour pigments** can be mixed (<10%) with resins, to help protect source polymers from light. **Antioxidants** are used to prevent oxidation, while **stabilisers** prevent polymer damage by long exposure to ultra-violet light. Additives such as **mica** improve electrical resistance, **asbestos** provides resistance to high temperatures and **graphite** reduces friction. Thermosetting plastics can be made stronger and tougher by reinforcing with other materials. Layers of paper and cloth in laminated form, glass fibre (GRP) and carbon fibres give strength and withstand strain and are used for specialist items such as sports equipment.

Available forms

A wide range of different powders, granules, pellets and viscous liquids provide the raw materials for processing into finished products. Manufacturers' standardised forms of plastics include film, sheet, tube rod and extruded mouldings. Many plastics are available in a variety of forms. **PVC** is used as a powder for dip-coating metals, as film when packaging and as sheet for vacuum forming. A wide range of **expanded plastics** and foams is available which are light in weight, good thermal and electrical insulators and provide flexibility with energy absorption. Gas pockets form a cellular structure giving either **open interconnecting cells** (sponges or foams) or **closed cellular structures** (expanded plastics). Polyurethanes are a good example of thermosetting polymers used in this way. In open cell form, polyurethane foam is flexible and used in upholstery sponges. However, polyester polyurethane foam has a closed cell structure, making it rigid and buoyant and suitable for internal reinforcement (e.g. in boats and aircraft wings). Expanded polystyrene is well known as a lightweight packaging material. Its volume consists of 90% air bubbles, making it a good thermal insulator with excellent flotation qualities (Figure 2.87).

Fig. 2.87 *Expanded polystyrene used in packaging*

Identification of common plastics

Most types can be identified by following basic procedures and using the workshop tests illustrated in Figure 2.88. Whether using just one or several samples, carefully record and analyse your findings. Do not rely on just one test!

Fig. 2.88 Identification of common plastics

⚠ Safety

All testing should be done under strictly controlled conditions.

Wear safety gloves and use tongs to avoid burning molten plastic.

Fumes can be poisonous, always ensure good ventilation.

1 Appearance

- Is the specimen (a) stock material (b) raw polymer (c) a finished product?
- If (c) try to establish the method of manufacture or processing techniques. Look for tell-tale flow lines, sprue or ejector pin marks.
- What is the colour? Is it pigmented, transparent, translucent or opaque?
- Try to establish the basic group (i.e. thermoplastic or thermoset).

2 Rigidity and feel

- How hard is the sample? Try to scratch it.
- How does it cut? Cleanly into slivers indicates a thermoplastic, if it flakes, powders or chips then a thermoset is indicated.
- Drop the sample on a hard surface. Styrenic thermoplastics (e.g.polystyrene) give a metallic ring, mouldings made from co-polymers and high impact material do not.

3 Bending

- Try to bend the sample and note the reaction and recovery.

4 Heating

- Heat a metal sheet over a bunsen flame or electric plate and place the sample on it. Does it soften, remain hard, flow, bubble, swell, or char? Thermoplastics soften rather than melt, thermosets do not melt. If the sample softens rather than hardens with continued heating, it indicates an uncured compound which has heat-cured.
- Try to describe the odour of any fumes, relate to familiar smells (e.g. burning rubber, sweet-fruity, marigolds, burning paper, acid, rancid butter).

5 Burning

- Hold a small piece in flame, when ignited withdraw it.
- Inflammability, does it burn? Is it self extinguishing?
- What is the colour of the flame?
- Note the type of flame (e.g. steady, short, tall, spluttering).
- Note the type of smoke (e.g. none, black, thick black, shot streamers).
- Is the burning accompanied by a distinct smell?

6 Density/specific gravity

- Put a sample into water. Stir to remove air bubbles. Does it float/sink?
- Specific gravity results can be found by relating weight to weight of water displaced. Remember to consider the presence of fillers and reinforcements.

	PVC	Polystyrene	Polyethylene	Phenolic	Acrylic
Test 1	Can be plasticised or uPVC	Various grades – toughened, transparent	Low or high density	Dark coloured	Clear, translucent and opaque
Test 2		Toughened			
Test 3	 Plasticised	Toughened	Low density / High density		
Test 4	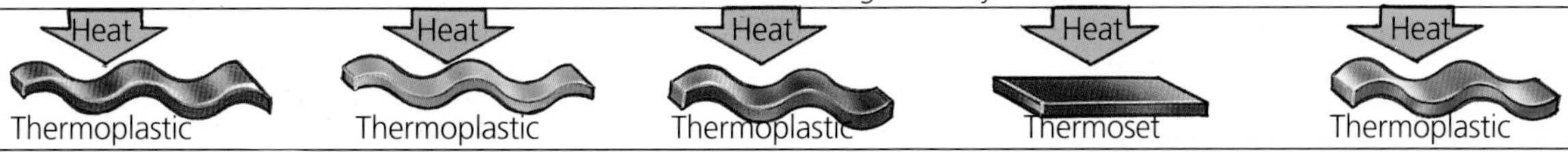 Thermoplastic	Thermoplastic	Thermoplastic	Thermoset	Thermoplastic
Odour	Pungent acrid	Like marigolds	Burning candle	Burning wood	Strong floral and fruity
Test 5	 Burns only with difficulty / Softens and chars at base	Melts, bubbles at edges	Plastic drips	Swells and cracks Self extinguishing	Bubbles and boils
Test 6	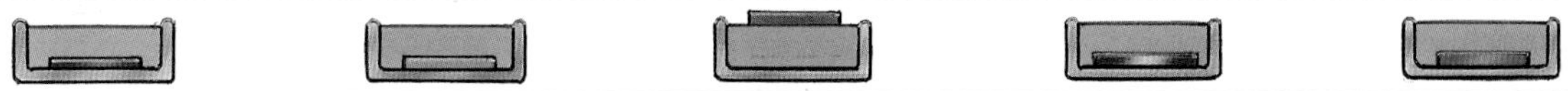				

a) Thermoplastics

KEY Suitability for project work ◆ poor ◆ ◆ average ◆ ◆ ◆ good

MATERIAL	PROPERTIES AND WORKING CHARACTERISTICS	USES
Polythene (polyethylene) (LDPE)	Low density: tough, common plastic good chemical resistance, flexible, soft, attracts dust, electrical insulator, wide range of colours	Detergent, squeezy bottles, toys, packaging film, carrier bags, TV cable
(HDPE)	High density: stiffer, harder, high softening point, can be sterilised, waxy feel ◆ ◆	Milk crates, bottles, pipes, houseware, buckets, bowls
Polypropylene (PP)	Light, hard, impact resistant even at low temperatures, good chemical resistance, can be sterilised, easily joined, welded, good resistance to work-fatigue, bending, hinges, good mechanically ◆ ◆ ◆	Medical equipment, syringes, containers with integral hinges, string, rope, nets, crates, chair shells, kitchenware, film
Polystyrene (PS)	(a) Conventional: light, hard, stiff, colourless, transparent, brittle, low impact strength, safe with food, good water resistance	Model kits, packaging, disposable plates, cups, utensils, TV cabinets, containers
	(b) Toughened: increases impact, strength, pigmented	Toys, refrigerator linings
	(c) Expanded/foam: buoyant, lightweight, crumbles, good sound/heat insulator ◆ ◆ ◆	Sound and heat insulation, packaging
Polyvinyl chloride (uPVC)	Good chemical, weather resistance, stiff, hard, tough, lightweight, wide colour ranges, needs to be stabilized for outdoor use	Pipes, guttering, bottles, shoe soles, roofing sheets, records, window frames
Plasticised (PVC)	Soft, flexible, good electrical insulator ◆ ◆ ◆	Underseal, hosepipes, wall coverings
Polymethyl methacrylate (Acrylic) (PMMA)	Stiff, hard, clear, very durable, IOX impact resistance of glass, but scratches easily, excellent light transmission, fibre optic qualities, safe with food, good electrical insulator, colours well, easily machined, polishes well ◆ ◆ ◆	Light units, illuminated signs, record player lids, aircraft canopies, windows, rear car lights/reflectors, furniture, sanitary ware
Polyamide (Nylon)	Creamy colour, hard, tough, resilient to wear, low co-efficient of friction, bearing surfaces, self-lubricating, resistant to extremes of temperature, good chemical resistance, machines well, difficult to join except mechanically ◆ ◆ ◆	Bearings, gear wheels, casings for power tools, curtain rail fittings, combs, clothing, stockings, hinges, filaments for brushes
Cellulose acetate	Tough, hard and stiff (can be made flexible), resilient, light in weight, transparent, non-flammable, easily machined, absorbs some moisture ◆	Pen cases, photographic film, cutlery handles, knobs, lids, spectacle frames, containers
Acrylonitrile butadienestyrene (ABS)	High impact strength and toughness, scratch resistant, light and durable, good appearance, high surface finish, resistant to chemicals ◆ ◆ ◆	Kitchen ware, cases for consumer durables (e.g. cameras), toys, safety helmets, car components, telephones, food processors/mixers

b) Thermosetting plastics

MATERIAL	PROPERTIES AND WORKING CHARACTERISTICS	USES
Urea-formaldehyde (UF)	Stiff, hard, strong, brittle, heat resistant, good electrical insulator, wide range of light colours, adhesive (Aerolite) ◆	(White) electrical fittings, domestic appliance parts (e.g.knobs), adhesives (wood), coating paper, textile
Melamine-formaldehyde (MF)	Stiff, hard, strong, scratch resistant, low water absorption, odourless, stain resistant, resists some chemicals, wide range of colours ◆ ◆	Tableware, decorative laminates for work surfaces, electrical insulation, buttons
Polyester resin (PR)	Stiff, hard, brittle (resilient when laminated GRP), good heat and chemical resistance, electrical insulator, resists ultra-violet light, good outdoors, contracts on curing, takes colour well ◆ ◆	Casting, encapsulation, embedding, panels (with GRP), boats, car bodies, chair shells, containers
Epoxy resin (epoxide) (ER)	High strength when re-inforced, good chemical and wear resistance, resists heat to 250°C, electrical insulator, adhesive for bonding unlike materials, low shrinkage ◆ ◆	Surface coatings, castings, encapsulation of electronic components, adhesives, laminating paper, PCB, tanks, pressure vessels

PLASTIC FINISHES

Excellent resistance to corrosion and decay makes protective surface coatings unnecessary for this group of materials. The scope for decoration, however, is almost unlimited. The high-quality finish associated with the **'self-finishing'** nature of the material is closely linked to the respective manufacturing process. In most cases the quality of the mould is directly reflected in the products' finish. The latter may be transparent, translucent or opaque and range from matt to high gloss, while also incorporating different **textures**. Colour and tone are easily changed through the addition of chemical dyes. Like most other materials it also takes paint finishes, as illustrated in Figure 2.89.

The finish on cut and shaped edges in acrylic need to match the other surfaces. Planing or drawfiling removes major scratches and blemishes. Final cleaning-up is undertaken with **wet and dry paper** (silicon carbide) of different grades (grits, 400–120). Acrylic is

Fig. 2.89 *A painted cast resin model*

polished with mild abrasive metal-type polishes or special **anti-static** creams to reduce the level of dust attracted to its surface. Alternatively, a buffing machine is used. **Vonax**, a compound formulated for plastics such as acrylic, is applied to a soft calico mop. Care must be taken to avoid permanent damage by overheating the surface. Acrylic is usually polished even before the process of line-bending.

Thermosetting laminates such as Formica are colour printed and provide hard, heat-resistant, hygienic and decorative surfaces, suitable for kitchen worktops.

Fig. 2.90 *Vinyl detail on a clock*

Materials like **vinyl** can be used with **CAD/CAM** (refer to Chapter 1) for graphic applications such as letters, numbers, labels and logos (Figure 2.90). Engraving with Cam 2 cutting equipment and techniques is also used to produce signs in a variety of laminated plastics (Figure 2.91). In industry, plastic containers have information or decoration applied directly by printing, or added later using plastic film transfers.

Fig. 2.91 *Industrial plastic sign*

MATERIALS FOR THE FUTURE

As new materials are developed design technologists must think about the implications of the changes. They have responsibilities in planning and extending all aspects of **recycling** materials. Nowadays products are increasingly being made from **composite** (more than one substance) materials. One product may comprise of wood, metals, polymers, elastomers and ceramics, each retaining and contributing its own special properties. It is important to record these properties and how they react in combination. For example, **carbon fibres** embedded in resin combine high tensile strength with low density and provide better corrosion resistance and fatigue performance than most metal alloys. The properties of most materials remain more or less

constant in use. **'Smart'** materials, however, respond to external factors such as difference in light or temperature levels and change in some way. They are now being applied to everyday products such as sunglasses, which darken as light intensity increases.

A **shape memory alloy (SMA)**, a mixture of nickel and titanium, called **Nitinol** can be made to 'remember' a shape as a result of special heat treatment. If bent at room temperature into a paper clip shape, it will stay bent. But if the temperature is raised to its **transition temperature** (70°C) it immediately straightens out – a cycle which can be repeated millions of times. Because SMA has a high electrical resistance, it can be heated to its transition

temperature by passing an electrical current through it. Its ability to provide large forces and movement at a precise temperature, makes it suitable for a variety of applications including electronic connectors, triggers and valves.

Other futuristic materials include unusual developments with **elastomers** (rubber). **'Anti-rubber'** expands if stretched and shrinks when squeezed, what is known as the **'negative poisson effect'**. The key to its bizarre properties apparently lies in its microscopic structure, which resembles irregular, buckled-in cubes. These re-entrant foams offer wide-ranging applications, from shock absorbers to filters, because their pores open as pressure increases, so they clog less easily.

Putting it into practice

1 What impact has the use of 'new' materials had on product design in the twentieth century? Illustrate your answer with suitable examples.

2 Why is the selection of materials an important design consideration? Use examples of products found in the kitchen to illustrate your answer.

3 Choose three materials that would be suitable for an outdoor modular climbing frame. Your answer should make reference to strength, durability, appearance and safety.

4 Compare the qualities and properties of two 'composite' materials.

5 Explain what you understand by the properties and characteristics of metals?

6 Compare and contrast the use of natural air seasoning with the kiln seasoning of timber – make references to 'moisture content'.

7 Describe how environmental changes (i.e, moisture humidity, sunlight etc.) effect wood, metal and plastic materials. How can these effects be minimised?

8 Define the terms 'hardness' and 'toughness' when applied to materials. Use sketches to describe how simple tests can be used in a school workshop to compare these properties.

9 With reference to material testing, explain each of the following: (a) elastic limit, (b) tensile testing, (c) hardness testing, (d) impact testing, (e) plastic extension and (f) necking.

10 Sketch and label three different types of wood-based manufactured board, stating their advantages and disadvantages when compared to solid timber.

11 Identify the main defects found in timber.

12 Draw up a table to suggest a suitable material and appropriate finish for each of the following products: garden gate, canoe, jewellery bracelet, candle holder, pull-along toy, house nameplate, kite, wheel-bearing, chess set, toolbox.

13 Explain the differences in the structure and properties of thermoplastics and thermosetting plastics.

14 By what means can different types of plastic be identified?

15 By means of sketches and brief notes, show the stages in the edge finishing of the following sheet materials: (a) acrylic, (b) Blockboard and (c) thin metal.

16 Explain the following terms applied to metal finishes: (a) anodising, (b) electro-plating, (c) enamelling, (d) etching.

By means of sequential sketches and brief notes show one of the above in detail.

17 Explain each of the following heat treatment processes applied to metal: (a) annealing, (b) Normalising, (c) hardening and tempering, (d) case-hardening.

Fig. 2.92

3· The forming processes

Manufactured products must be made to the highest standards and be as cost effective as possible. This requires taking into account the relationship between materials, form and manufacturing processes. When you are designing a product, it is important to consider early on how it will be made . You should do this as you develop your chosen idea. To do this you need to build up an awareness of materials by working with a wide range of them and exploring different techniques and processes.

This and the following chapter will help you to develop an overview of the different manufacturing processes and help you to decide upon the most appropriate forming processes to use.

Fig. 3.1 *Working with GRP*

Fig. 3.2 *Working with wood*

The basic forming processes used in modern manufacturing are **wasting** (i.e. taking something away from the material), **reforming** (e.g. changing the state of a material, from solid to liquid to solid, as in **casting**) and **deforming** (changing the shape of a material without wasting). **Fabrication**, the process of joining materials together, is examined in the next chapter.

All of these processes can be adapted to suit the needs of particular manufacturers and the types of material being used. For instance, the blacksmith, a traditional worker in metal, is able to produce complex shapes by using a combination of the flat and curved surfaces found on the anvil, with a variety of tools (deforming). The flexibility which this offers can be adapted to meet many different requirements.

Looking at each forming process in detail, this chapter will cover both the use of hand tools and craft techniques, as well as the sophisticated machine tool technology used in industry. In industry especially, using tools is a very important way of increasing efficiency. For instance, the manufacturing techniques used to produce curved panels for the aircraft and motor industry (such as those in Figure 3.4) require complex and expensive tooling. Powerful presses are used to force the material into shape and produce quantities of identical components.

Fig. 3.3 *Working with metal*

Although you will be working with different materials, many of the techniques you use will be the same – **marking out**, **measuring**, using **holding devices**. They will all be covered in this chapter. The most important common factor, **safety**, must be considered first.

Fig. 3.4 *Manufacturing*

The Health & Safety Executive (HSE) was created by the Health and Safety Work Act 1974. Its aims are to promote compliance with the Act, and thereby protect the health, safety and welfare of employees and to safeguard others, principally the public, who may be exposed to risks from work activities.

Their three main areas of activity involve:

HSE
Health & Safety Executive

1 Inspection – HSE inspects workplaces and enforces legislation by means of informal advice, written advice, or formal enforcement action (e.g. prosecution).

2 Guidance – HSE publish guidance on a range of subjects to tell employers how to comply with the legislation. It may be specific to the health and safety problems of an industry or of a particular process used in a number of industries.

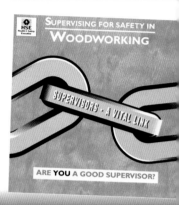

WORK SAFELY

When you are 'realising' (making) your designs it is most important to make them safely, using safe working practices in a safe working environment. The single most common cause of accidents in workshops is human carelessness, and the most effective way of creating a safe working environment is to behave in a mature and responsible manner. You must always think about your own safety, and that of others – one brief lapse of concentration could result in an accident that could change your life, or somebody else's life forever. Always follow instructions and safety rules carefully.

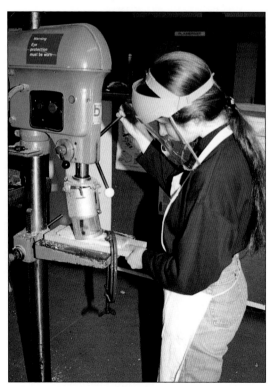

Fig. 3.5 *Observing safe working practice – this student is using a pillar drill with chuck guards in place*

Check

Fig. 3.6 *Safety check symbol*

This safety symbol is used throughout the book to draw your attention to matters you need to know, potential hazards, and warnings you need to heed, relating to **safety**.

What you wear

■ Before starting work with any tools or equipment, especially machines, remove or tuck in ties, take off jewellery and tie back long hair.
■ Wear aprons or overalls to protect clothes, as well as stout shoes.
■ Wear special protective clothing and eye protection when required
■ Always wear the safety items provided for you and make sure they are put back for others to use.

Fig. 3.7 *Eye protection symbol*

How you behave

■ Behave sensibly at all times. Do not run, shout, fool about, or distract others.
■ Move around and carry tools and materials in a safe manner.

Modernisation of legislation before the 1974 Act – this includes the production of Approved Codes of Practice, which have a special legal status and employers can be found at fault if they have not followed the relevant procedures.

Health and Safety regulations cover a wide field – ventilation, temperature, lighting, cleanliness and waste materials, maintenance, safety devices, noise levels, special clothing, and so on. Some standards are absolute, for example 'all mines should have two exits'. Some activities or substances are so hazardous that they have to be licensed, for example, asbestos removal and explosives.

The HSE ensure that risks to people's health and safety are controlled in ways that are proportionate to those risks, that allow for technological progress, and that take into account the cost to the employer.

Health and Safety Inspectors have the power to enter premises, take measurements, photographs, recordings and samples as part of an investigation.

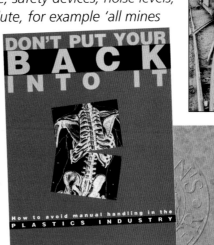

Know how – accident procedure

It is important that you know what to do in the case of an accident or fire.

- Tell the teacher (or responsible adult) immediately when any accident occurs.
- Have all injuries, however slight, properly attended to.
- Know the location of First Aid boxes.
- Be calm, helpful and try to avoid panic.
- Know where to find the Emergency Stop buttons.
- Always keep gangways clear and emergency exits accessible.

Fig. 3.8 *Safety symbols – be aware of them*

Safe working practice

- Keep your work area well organised, clean and tidy.
- Keep areas between work benches and around machines clear to avoid potential hazards to trip, or fall over.
- Check the condition of your tools and immediately report any which are blunt, broken or have loose parts, etc.
- When using sharp-edged tools remember that both hands must be kept behind the cutting edge.
- Clean all equipment thoroughly after use and return to its proper place. Report any faults, damage or breakages.

A safe working environment

Always be aware of the potential hazards in the workshop – electrical, chemical, heat and dust.

- Check the condition of portable appliances (e.g. soldering irons). Ensure they are electrically and mechanically safe (e.g. no frayed or burnt flex, no loose screws).
- Leave hot tools, or hot pieces of work in a safe place to cool before putting them away. Inform other people by using a clear warning sign that they are HOT!
- Always read instructions on chemically based products (e.g. glues and solvents). Always heed the warnings (e.g. use in a well ventilated room).
- Consult your teacher about the disposal of all chemical waste, including paper towels and cloths.
- Avoid inhaling large amounts of dust – wear a face mask when using power sanders.
- Clean up and wash hands thoroughly after any work.

Fig. 3.9 *More safety symbols – warnings in yellow and black, mandatory instructions in blue and hazard warning diamonds*

MARKING OUT

To achieve quality products when working with materials you need a high degree of accuracy when measuring, attention to detail and the ability to work to fine tolerances (see page 28).

Templates

If you need to create irregular shapes, it is best to make a paper or card template which can be glued directly on to a surface (use an adhesive that allows the paper to be easily removed) and cut round.

If you are working with a sheet plastic that has a protective film of paper over it, it may be necessary to remove the film before you attach your template.

Templates are especially useful when you need to cut out the same shape a number of times. Use any thin gauge material to make templates for this (e.g. plywood, aluminium, acrylic) and draw around it.

Marking out directly on the surface of materials

For **regular shapes**, measuring and marking out needs to be done from a flat face (straight edge). In the case of metals and plastics these are referred to as **datum surfaces**. These are the points from which you always need to start your measuring in order to avoid cumulative errors (Figure 3.11).

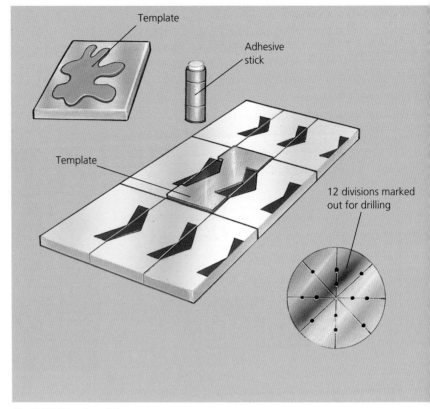

Template

Adhesive stick

Template

12 divisions marked out for drilling

Fig. 3.10 *Using templates*

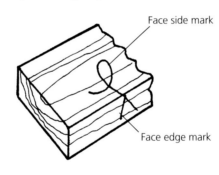

Face side mark

Face edge mark

Fig. 3.12

With timber it is necessary to create these surfaces – by selection, in the case of pre-planed wood, or by planing, with rough sawn timber. The surface which is created first is commonly referred to as the **face side**, and the other is known as the **face edge**. They are marked accordingly, using a pencil, with a loop leading to a cross (Figure 3.12).

Marking light points or lines, which can easily be corrected or removed, is usually done with a pencil on wood, and a chinagraph pencil or fine-line spirit pen, on plastics. The more accurate the work needs to be, the finer the line should be. A **marking knife** or craft knife leaves a permanent cut in wood, usually across the grain, severing the fibres

and making a clean edge. A **scriber** is used to produce light scratch lines on metal and plastic surfaces. On metal, it is advisable to first coat the immediate area with a fast drying coating of 'engineer's blue' so that the marks you make will be visible (a dark spirit-based felt-tip pen will work just as well).

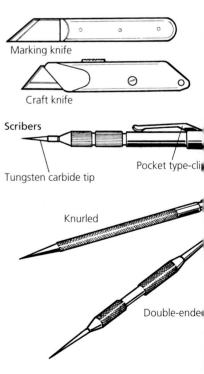

Marking knife

Craft knife

Scribers

Tungsten carbide tip

Pocket type-clip

Knurled

Double-ended

Fig. 3.13 *Marking out tools*

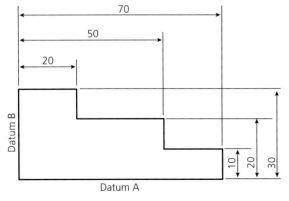

Datum B

Datum A

70

50

20

10

20

30

Fig. 3.11

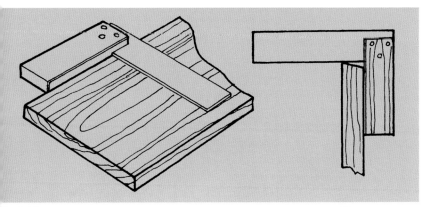

Fig. 3.14

A **try square** is used to make lines at 90° to an edge. It also serves to check that edges have been cut at right angles to one another. This can be tested by holding both up to the light (see Figure 3.14). If a wedge of light is visible, the edge is not square.

A **marking gauge** is used on wood to mark straight lines parallel to an edge. This tool has an adjustable stock and is set using a rule (Figure 3.15). In use, the stock must be kept pressed tightly against the face edge. It is pushed away from the body, with the spur inclined, which leaves a small groove along the grain. Similarly, a **mortise gauge**, which has two spurs, makes two parallel lines. A **cutting gauge**, which has a knife blade, is used to make cuts across the grain from a prepared end.

Setting a marking gauge

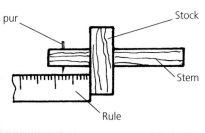

Spur · Stock · Stem · Rule

Fig. 3.15 *Using gauges for wood*

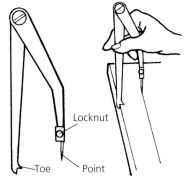

Locknut · Toe · Point

Fig. 3.16

To mark out lines parallel to an edge on metal, **odd leg calipers** are used (Figure 3.16). The stepped foot must be kept firmly pressed against the edge of the work.

A more accurate method, especially useful for marking parallel lines on curved surfaces (such as cylindrical metal or plastic tubing) is achieved by using a **scribing gauge** on a **surface plate** together with a **vee block** (Figure 3.17).

Surface plate

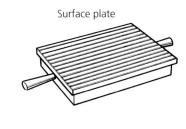

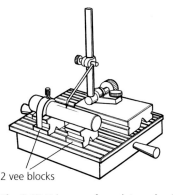

2 vee blocks

Fig. 3.17 *Using a surface plate and scribing block to mark lines on a metal tube*

Marking holes

Compasses or **dividers** are used to mark out circles. Those that have a screw for adjusting the radius give the greatest accuracy. When a hole is to be drilled in wood, it is first positioned with a pencil cross, and a **bradawl** is then used to make a small hole to locate and guide the drill. Small indents in metal also need to be made to locate the exact centre and prevent drills from sliding about. This is done by making a 'dot' with a **centre punch**, positioned where two scriber lines meet.

To help to cut out curved lines or arcs on metal a **'dot' punch** is used at regular intervals along the cutting line.

It is always advisable to mark out and **check** measurements carefully, **before cutting** out. Also, identify waste by scribbling on it. T his can be very helpful, especially when it's time to remove it.

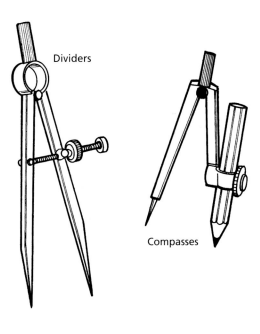

Dividers · Compasses

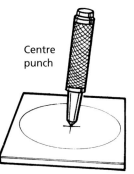

Centre punch

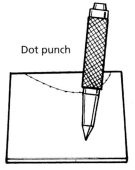

Dot punch

Fig. 3.18 *Tools for marking holes and curves*

UNITS OF MEASUREMENT

The preferred unit of measurement used in design & technology is the millimetre. When it is written down or shown on drawings, it is abbreviated to 'mm'. For example, 2 centimetres is expressed as 20 mm and 2 metres as 2000 mm. Measurements of less than one millimetre are shown as a decimal preceded by a zero. For example, half a millimetre is shown as 0.5 mm.

MEASURING INSTRUMENTS

The measuring instrument you choose is determined by the degree of accuracy required, not by the type of material you are measuring. In most situations a **steel rule** will be adequate. However, if a high level of accuracy is required, a precision instrument, such as a **Vernier caliper** or a micrometer may be more appropriate. Measuring instruments are used not only to make and verify the dimensions, but

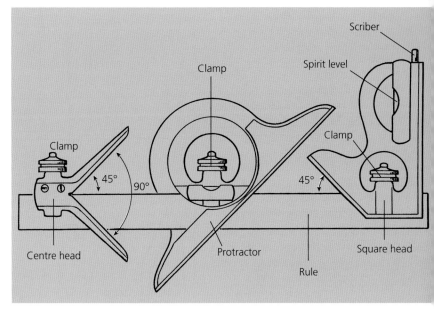

Fig. 3.19 *A combination square*

also to measure the accuracy and relationship of surfaces (such as flatness, squareness, parallelism and roundness). To ensure that your measurements are accurate you must use and maintain these instruments properly.

One of the most versatile measuring instruments is a combination square, which includes a rule, square head, centre head and protractor head, and can be used to meet most needs.

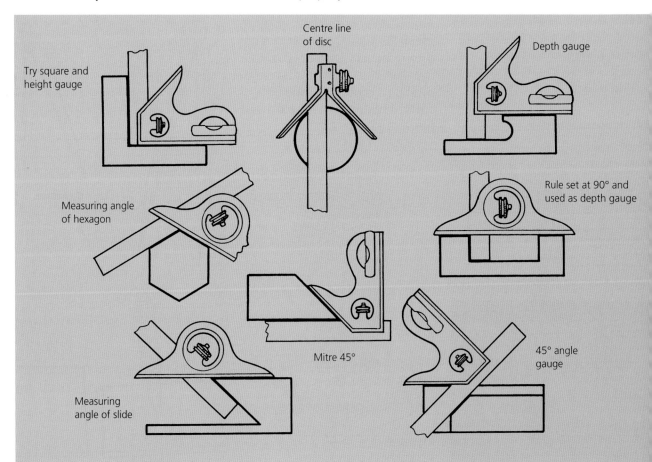

Fig. 3.20 *Using a combination square*

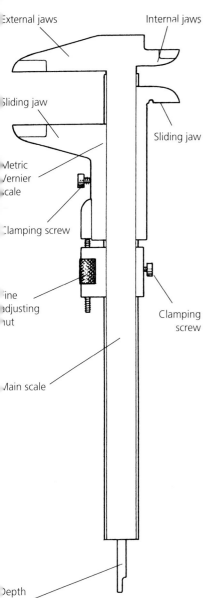

External jaws

Internal jaws

Sliding jaw

Sliding jaw

Metric
Vernier
scale

Clamping screw

Fine
adjusting
nut

Clamping
screw

Main scale

Depth
gauge

Fig. 3.21 Vernier caliper

For precise measurements, a Vernier caliper is often used (Figure 3.21). The scale that it uses, named after its inventor, sub-divides the main scale into smaller parts, and gives an accuracy to 0.02 mm. The Vernier scale is also applied to depth gauges and protractors.

Fig. 3.22

A **micrometer** (Figure 3.22) is specially designed to be used either as a comparitor, or for taking small measurements. The thimble has 50 equal divisions, providing an accuracy of 0.01 mm. To use a micrometer you turn the thimble to close the gap around the object you are measuring until it is held in place. Readings are taken by adding the visibly exposed whole numbers to any $\frac{1}{2}$ millimetres and then adding on the reading from the thimble (0–0.49 mm). This principle applies to micrometers of all types – external, internal and depth.

Figure 3.24 shows an **electronic digital micrometer**, which provides a precise direct readout. It can also be zeroed at any point, which enables its use as a comparitor.

Fig. 3.23 Using a micrometer

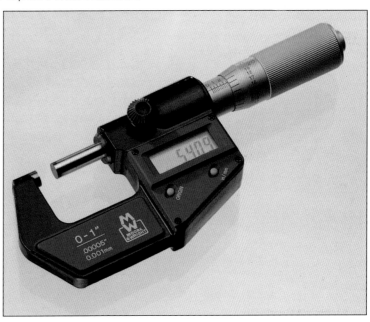

Fig. 3.24 Electronic digital micrometer

HOLDING DEVICES

Working with materials involves many operations, such as cutting and drilling, where the forces acting upon the material can be considerable. It is important, therefore, that materials are held firmly and securely to prevent any damage.

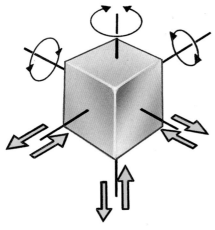

Fig. 3.25 *Movement can occur in a number of planes and directions simultaneously.*

Movement is restricted in one of two ways. Either by positive location (e.g. placing the material against a solid object) or by frictional resistance (e.g. being gripped by a vice).

Most **vices** are made of cast iron and are bolted to benches when in use. Their steel screw thread can exert a considerable holding force (approximately 1 tonne). An **engineers' vice** has hardened steel jaws that can be replaced. Aluminium or fibre clip-on vice jaws are also available to protect the work (Figure 3.27).

Fig. 3.26 *An engineers' vice*

Fig. 3.27 *Fibre clip-on vice jaws*

A **woodworking vice** can have a quick release mechanism, which allows the jaws to be opened and closed rapidly. They are fitted with replaceable hardwood (usually beech) cheeks.

To hold work for drilling or other machining operations, use a **machine vice**. There are many types and sizes available. Depending upon the torque (rotational forces) being exerted on the workpiece, the vice will need to be bolted down, or securely clamped to the machine table (Figure 3.29).

Fig. 3.29 *Machine vices*

A **bench hook** (sawing board) provides positive location against the cutting forces of a tenon saw, while a bench stop does the same against the cutting forces of a plane.

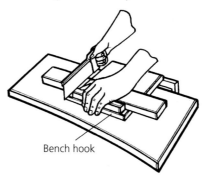

Bench hook

Fig. 3.30 *Using a bench hook with a tenon saw*

Fig. 3.28 *A woodworking vice*

Certain benches, including the portable D.I.Y. 'Workmate' (Figure 3.31) have adjustable dogs which can be used to hold different sized workpieces. In the past, this was only possible with the help of a **bench holdfast**. This is a single leg device which fixes into a metal reinforced hole in the bench/table top. By tightening the screw, the leg is twisted at an angle in the hole, thereby locking the workpiece solid against the top (Figure 3.32).

Fig. 3.31 *Workmate*

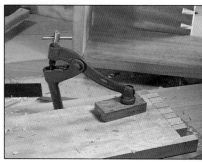

Fig. 3.32 *Bench holdfast*

Holding thin sheet material

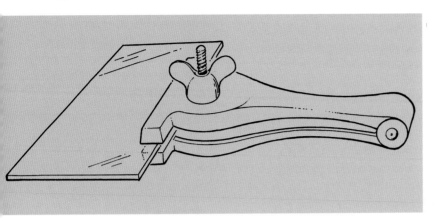

Fig. 3.33 *Holding thin sheet material with a hand vice*

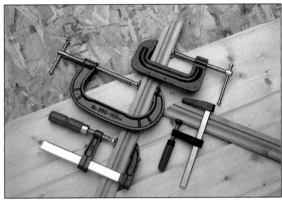

Holding thin sheet material can be a problem, especially if you want to drill holes using a pillar drill. Using a **hand vice** is one possible solution. Alternatively, you can use a **jig** (Figure 3.34). A jig is a holding device that is designed to be used when making more than one identical component, so it is especially helpful when you need to use a machine tool to perform several similar operations.

If you need to hold materials of variable thicknesses, build a jig that incorporates **toggle clamps**. Because of their action (Figure 3.35) toggle clamps are quick to use. Attaching toggle clamps to a tee piece, which can in turn be held in a machine vice, combines holding power with the versatility of quick release (Figure 3.36).

Fig. 3.37 *A selection of holding devices*

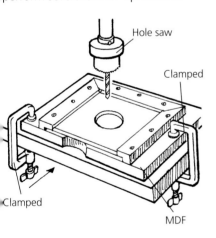

Fig. 3.34 *Using a jig for drilling*

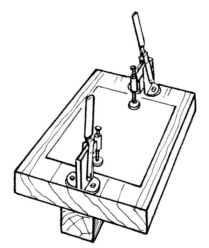

Fig. 3.36

Clamp off

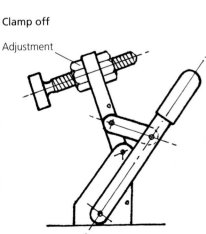

Fig. 3.35 *The holding action of a toggle clamp*

Clamp on

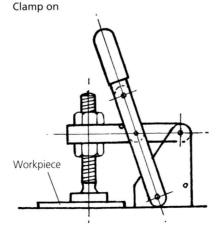

It is often necessary to hold pieces of material together when marking out or gluing up, as well as when cutting and drilling. Figure 3.37 shows a selection of the many holding devices available. **Sash cramps**, which are available in various lengths, are particularly useful for clamping longer lengths of wood. They give an excellent square pull when used with suitably shaped clamping blocks. **G-cramps** are excellent for general purpose fastening and are also available in a wide range of styles and sizes. **Toolmakers clamps**, made from case hardened steel, are used mainly for holding together pieces of metal. The two screws enable the jaws to open and close parallel to one another.

For more difficult holding problems there are various specialist frame, webb and corner clamps, which employ various tensioning devices.

WASTING

The process of taking something away from a piece of material is described as **'wasting'** – the piece that is removed is usually referred to as 'waste'. Using this forming process you create a new form or component by removing or cutting away any surplus material. This can be achieved by several means, such as electrical, chemical (etching) and thermal. The most common method, however, is **mechanical wasting**, using a range of cutting tools which are either hand or machine operated. All cutting tools must be made from a harder material than that being cut. Usually they are made from some form of metal alloy (e.g. carbon steel). In industry some harder materials such as ceramics and diamonds are also used.

The same **cutting action** is used in all wasting processes. It can be likened to driving a wedge into the material, causing the waste to split off. Tearing of the material occurs in front of the cutting edge, which then cleans up the torn surface (Figure 3.38). The **cutting angle** is important – tools must not rub the surface of the material because this causes unnecessary friction and heat. To avoid this, a **clearance angle** is created by lifting the back of the wedge part of the tool. To prevent further friction, **secondary** clearance angles are sometimes needed.

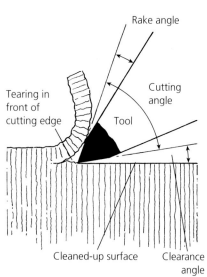

Fig. 3.38 *The cutting action*

The actual shape of the cutting tool and cutting angle depends upon the work it has to do and the material to be cut. It is important to keep your tools sharp as well as using them at the correct angles. This will ensure that efficient cutting takes place, leaving a good quality finish on the cut surface. If you are using machine tools, the speed of the cutting action is also critical.

Chiselling

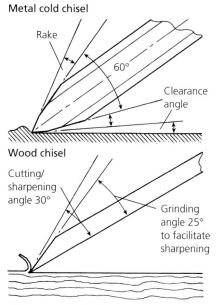

Fig. 3.39 *Using chisels for metal and for wood*

The chiselling action is used for both wood and metal. It is, essentially, the basic wedge cutting action. The wood chisels illustrated are used for paring and chopping.

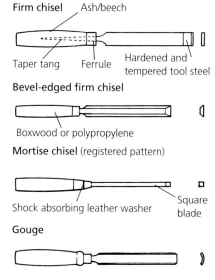

Fig. 3.40 *Wood chisels*

Paring is the term used to describe the removal of small shavings using hand pressure. It can be done either horizontally or vertically.

Chopping is the action of driving a chisel using blows from a mallet. It is most appropriate for removing large quantities of waste (e.g. making rectangular mortise holes).

A **plane** works very much like a chisel held at a particular angle. When you are planing it is important to keep the tool level by pressing down on the front and pushing from the back. Some hardwoods can be difficult to plane – closing up the mouth will reduce the risk of tearing. There are a variety of special planes available – plough planes for making grooves, rebate planes, spokeshaves for working curved surfaces and shoulder planes for cutting end grain.

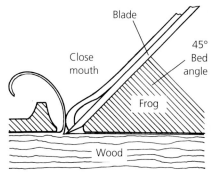

Fig. 3.41 *The cutting action of a plane*

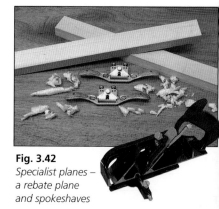

Fig. 3.42
Specialist planes – a rebate plane and spokeshaves

Filing

Filing is a versatile wasting process. Files are made of high-carbon steel which is hardened and tempered. The cutting is done by rows of teeth which remove small particles of the material, called 'filings'. Files are classified by their **length, shape** and **cut** (Figure 3.43). General purpose files are double cut, with small diamond-shaped teeth. Single-cut files are cut in one direction only and are most suitable for light finishing work. There are several grades of cut available for different tasks – **rough** and **bastard cuts** for coarse work, **second cut** for general use, and **smooth** and **dead smooth** for finer cuts prior to finishing. **Warding files** are used for thin narrow slots. **Needle files** ('Swiss files') are smaller precision versions of engineering files with dead smooth cuts. **Dreadnought files** have a curved cut suitable for the rapid removal of soft or fibrous materials (e.g. aluminium, copper, GRP).

Drawfiling is a finishing technique that is used after removing waste by crossfiling. A good quality surface finish can be obtained by leaving the scratch lines running in one direction along the workpiece.

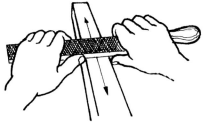

Fig. 3.44 *Drawfiling*

Rasps are similar to files, but have coarser, individual teeth and are more suitable for use on wood.

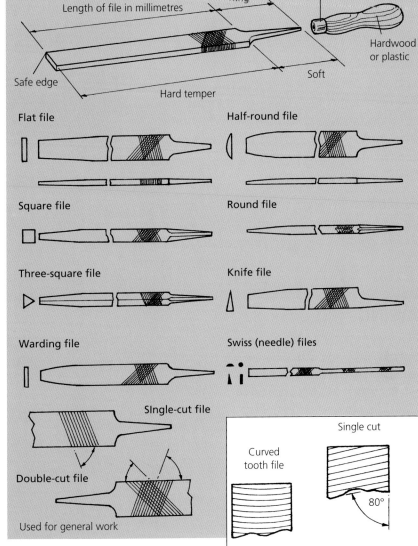

Length of file in millimetres

Tang

Ferrule

Hardwood or plastic

Safe edge

Soft

Hard temper

Flat file

Half-round file

Square file

Round file

Three-square file

Knife file

Warding file

Swiss (needle) files

Single-cut file

Double-cut file

Used for general work

Fig. 3.43 *Types of file*

Curved tooth file

Single cut

Double cut

80°

60°

80°

Enlarged cross-section of file

Direction of stroke

Work

Just enough pressure to make the file cut

Backwards and forwards movement

Material

Teeth cut on the forward stroke

Surform tools are an example of the evolution of traditional tools to meet modern requirements. They are available in a range of shapes and sizes and have replaceable blades (Figure 3.46). Each blade has hundreds of cutting edges with a hole to clear waste and avoid clogging. Blades are graded – **standard cut** for the rapid shaping of softwood and soft plastics such as nylon, **fine cut** for finishing hardwoods, acrylic and soft metals, and **special cut** for mild steel and plastic laminates.

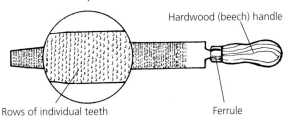

Hardwood (beech) handle

Rows of individual teeth

Ferrule

Fig. 3.45 *A rasp*

Fig. 3.46

Roy Walker and Bob Preece are specialist joinery manufacturers. They converted their premises into workshops from old farm buildings that had been derelict for many years.

The work that they carry out tends to be 'one off' and very specialised. Their most common contracts are to fit kitchens, bedrooms, shops and offices. Amongst their more unusual contracts was to convert a Grimsby fishing trawler into an oceanographic study vessel.

Roy and Bob still use many traditional joinery techniques, but with the large range of new materials now available (particularly manufactured boards) modern manufacturing techniques are used a great deal.

SAWING

There are many saws available, in numerous sizes and with different shaped and sized teeth that have been pitched to suit the material to be cut. The **pitch** is the number of teeth per 25mm – the more teeth there are, the finer the pitch.

In general, soft materials require a coarse pitch and hard materials a fine pitch. However, the main requirement is that as many teeth as possible are in contact with the workpiece at all times (Figure 3.47). For instance, for tubes and thin sectioned material a fine pitch should be used.

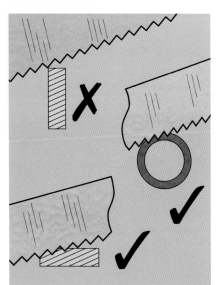

Fig. 3.47 *Use a saw and a saw action that keeps as many teeth as possible in contact with the material*

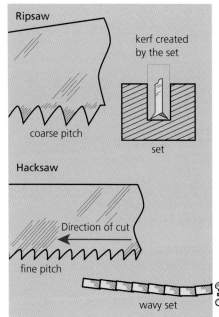

Fig. 3.48

The width of cut that a saw makes is called the **'kerf'**. This needs to be wider than the blade to avoid it jamming or getting stuck. To create the necessary clearance, the teeth are **'set'** by bending them out alternately to the left and right. Hacksaw blades are often set in the form of a wavy cutting edge. Figure 3.48 shows the blades and set of a ripsaw and a hacksaw.

Saws generally are classified by the materials they cut as well as the operational tasks they perform (Figure 3.49).

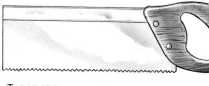

Tenon saw
(general purpose – wood)

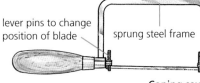

lever pins to change position of blade

sprung steel frame

Coping saw
(curves in wood or plastic)

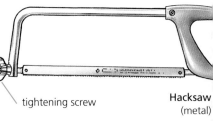

tightening screw

Hacksaw
(metal)

Junior hacksaw

Fig. 3.49

DRILLING

To make holes in most materials, including wood, metal and plastics, **twist drills** are used. The better quality, harder wearing type are made from high-speed steel (HSS). Those made from tungsten carbide can drill brick, concrete and, with special care, glass and ceramics.

For instance, when machine cutting laminated board the surface material tends to tear on the bottom surface as the blade passes through it. To solve this problem at Walker & Preece they use a machine called a 'machine panel saw' – a twin bladed circular saw especially for laminated board (see the photograph on the right). It has a small 'scribing' blade that cuts the bottom surface just before the larger saw blade cuts the board. This results in a cleanly cut surface on both sides.

Over the past 40 years saw blades have changed a great deal and there is now a great variety. Whereas 40 years ago High Speed Steel (HSS) circular saw blades were the best available, since then, to cope with new materials such as chip board, laminated boards and MDF, saw blades have been developed with Tungsten Carbide Tips (TCT) and Polycrystalline Diamond (PCD) tips.

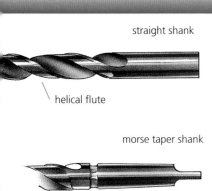

straight shank

helical flute

morse taper shank

Fig. 3.50 *Twist drills*

Drilling employs the basic wedge type of cutting action, with two cutting edges working together. The twist ('helix') forms flutes that carry the waste ('swarf') out of the hole (Figure 3.50). Smaller sized drills have **straight shanks** and are held in drill chucks. Larger drills (more than 13mm in diameter) often have **morse tapered shanks** that fit directly into machine spindles or lathe tailstocks.

When you need to make holes in wood that are more than 8mm in diameter, it is best to use a **boring bit** that has been designed for cutting wood (Figure 3.51). Most of these have spurs to sever the wood fibres cleanly. A centre screw pulls the bit into the wood and the helix (sometimes known as the auger) clears the waste from deeper holes. **Centre bits** are only suitable for shallow holes. **Forstner bits** have a point which guides the outer cutting ring, instead of a centre screw. They are used to produce clean, flat bottomed holes.

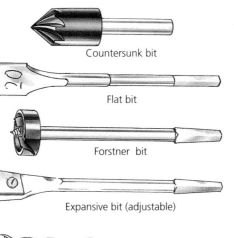

spur   cutter

screw point

Fig. 3.51 *Wood boring bits*

Centre bit

Countersunk bit

Flat bit

Forstner bit

Expansive bit (adjustable)

Jennings pattern auger bit

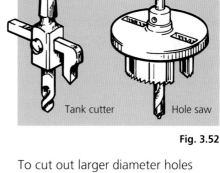

Tank cutter   Hole saw

Fig. 3.52

To cut out larger diameter holes from thin sheet material use adjustable **tank cutters** or a **hole saw**, which has interchangeable toothed cutting rings ranging in diameter from 20–75mm (Figure 3.52).

A **hand drill** can be used for smaller drills. It has a limited chuck capacity, but it can be used horizontally as well as vertically. A **carpenter's ratchet brace** is designed for square-shanked bits.

Fig. 3.53 *Hand drills and carpenters ratchet braces*

*Some people have long held the view that traditional joints are essential for quality woodwork. However, with modern machines, assembly techniques and adhesives, 'comb joints' are both quicker to produce and stronger. At Walker & Preece many joints are cut using **machine tools**. Machine-cut joints are very accurate. In the photographs you can see a machine-cut mortise and tenon joint and the machine for cutting tenons being used.*

*Most modern kitchens have fitted work surfaces, and increasingly, these have jointed corners. The **power tool** that is used to make these is called a 'router'. First of all, the profile of the corner is machined with the router following an aluminium jig (see right). The router is then used to cut out pockets for special clamps, and a 'biscuit cutter' is used to cut an accurate slot in the edge of the work surface along the joint face.*

POWER TOOLS

Using tools that are powered by electricity or compressed air can make the process of wasting much easier and quicker. Depending upon the nature of the tool, they can be hand-held (e.g. drill) or bench-mounted (e.g. bandsaw). Many electrically powered tools are available that operate on domestic mains electricity, but factories and construction sites, in the interests of safety, often only allow the use of low voltage tools that are powered from a mains operated transformer.

Compressed-air operated tools are used when electricity is not available directly or when the working environment prevents its use (i.e. if there is a risk of igniting flammable liquids or gases or the workplace is damp or wet). Provided that they are used correctly, both electrically operated tools and 'air tools' are safe to use.

Portable power drills have varying degrees of sophistication, including different chuck sizes, depth stops and variable speeds, as well as a hammer ('percussion') action for masonry.

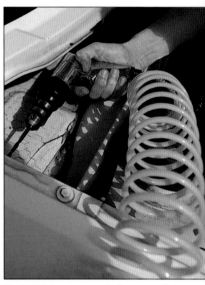

Fig. 3.55 *A compressed-air operated power drill in use*

Fig. 3.54 *A range of different power tools*

> ⚠️ **Safety note**
>
> Take care when using any power tool. They have sharp tools or blades driven by a powerful motor. Use any guards that are fitted to machines, and remember to wear eye protection. Do not put compressed air pipes or nozzles close to the skin.

The 'biscuit' is a thin piece of beech that is glued in when the corner is assembled in order to act as a key and to accurately line up the top of the work surface (see right).

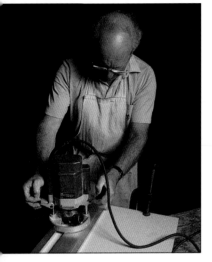

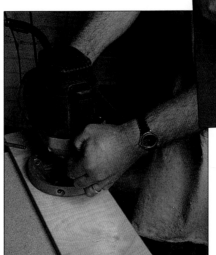

Cutting pockets for clamps

Biscuit cutter

MACHINE TOOLS

The larger power tools are usually fixed to the floor or a bench and are known as 'machine tools'. There are many different machine tools used in modern manufacturing, ranging from simple drilling machines, lathes and milling machines through to complex computer controlled machining centres.

One of the most common examples of a simple machine tool is a pillar drill, which can be floor- or bench-mounted (Figure 3.56). It has a removable Jacobs chuck so that larger sizes of drills with tapered shanks can be fitted. To use a pillar drill you lower it on to the work using its rack and pinion mechanism, which usually has a depth stop to enable you to drill holes to a precise depth.

When you work with any machine tool, it is vital that you use the correct cutting speed. As a general guide when drilling, the harder the material and the larger the drill, the slower the **cutting speed**. Soft materials and small diameter drills require higher speeds.

Fig. 3.56

⚠ WORK SAFELY WITH MACHINE TOOLS

Always follow this check list when working with machine tools:

1 Always ask permission from your teacher before using any machinery and electrical equipment.

2 Always get instructions and be shown how to operate the machine in question.

3 Never display false confidence. If you are unsure, ask to be shown again.

4 Wear eye protection whenever appropriate.

5 Know where the ON/OFF switches and emergency stop buttons are.

6 Ensure that the machine is set correctly and that the work piece is secure.

7 Always check that the guards are correctly positioned before starting the machine.

8 Never use machine tools if you are in a workshop on your own. Never leave working machine tools unattended – always switch them off when you have finished.

9 Do not distract others, or allow yourself to be distracted.

10 Always clear up properly and put away specialist equipment – leave it as you would wish to find it!

Wood lathe

The wood lathe can be used for turning in two ways – **on the headstock** or **between centres**. Either side of the headstock is suitable for turning using a face plate to produce such 'flat' products as bowls, dishes, formers and bases. To handle longer pieces of work, such as legs and spindles, you need to turn between centres. To do this, a **drive centre** is pushed into the spindle to turn the work which is supported at the other end by a **'dead centre'** located in the tailstock. A **3-jaw chuck** (see page 100) attached to the internal spindle can offer additional drive facilities for small diameter work.

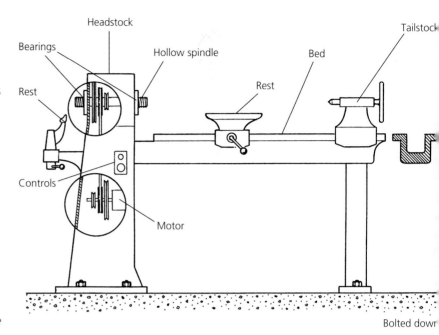

Fig. 3.57 A wood turning lathe

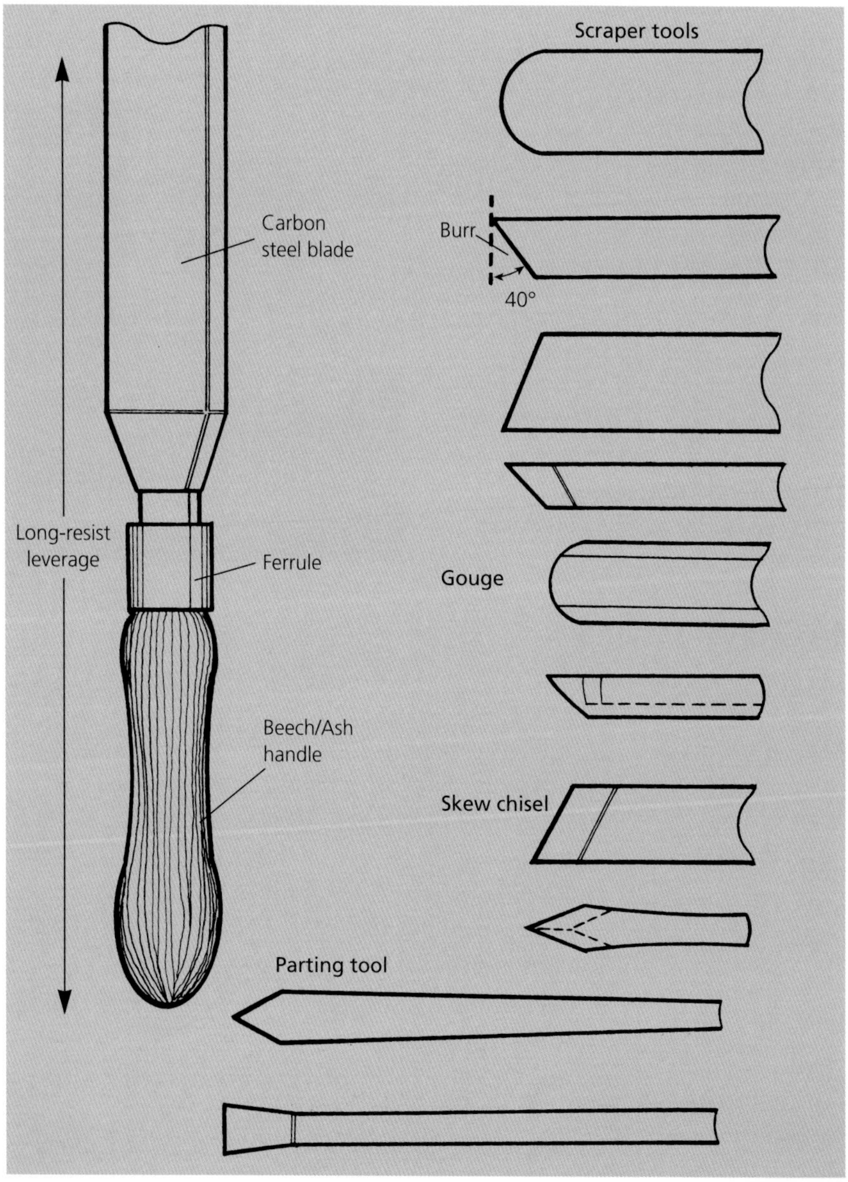

Bolted down

Fig. 3.58 Lathe tools

While the lathe holds the work and turns it, the cutting is carried out using a range of special lathe tools, held firmly. The long hardwood handles of the tools provide the leverage needed to withstand the turning force. Thick heavy **scraper tools** are the easiest type to use, and they give good results if kept sharp. A 'burr', which does the actual cutting, is produced on the grinding wheel. In use, the scraping action produces shavings, and frequent resharpening is necessary. The shape of each tool varies (i.e. curved or straight – see Figure 3.58). For the initial stages of turning, a round nosed tool is the best one to use.

Gouges and chisels use the more familiar cutting action. They too must be kept sharp, using an oilstone, so that the burr is removed to give the wedge shape. **Gouges** can remove quantities of material quickly. Therefore they are suitable for truing-up and rough shaping. **Chisels** are more suited to final shaping and fine finishing. However, they are not easy to use and require a lot of practice. The **parting tool** is used, as its name implies, to cut down, part-off, or remove material. It is taper ground, with the cutting edge wider than the blade.

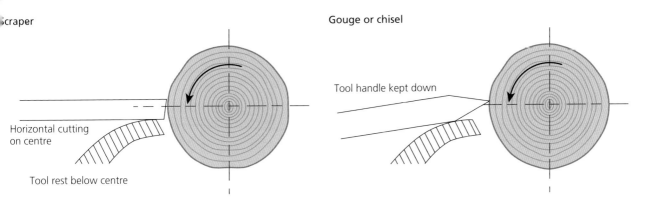

Fig. 3.59 *The cutting actions of lathe tools*

The height of the **tool rest** on a wood lathe is adjusted to accommodate the different approach angles required by each tool. Cutting takes place on or just above centre (Figure 3.59).

Proper preparation is important. If you are turning on a face plate (right-hand thread for internal, left-hand thread for external) removing corners from planks or sawing into a cylindrical form is essential. A spacing disc is also needed for some shapes. This is glued to the workpiece with a paper membrane in between to make for easier separation once the work is completed. The **face plate** needs to be attached to the work with thick gauge (i.e. No. 10) wood screws which need to be long enough to hold the work securely, without interfering with the required shape (Figure 3.60).

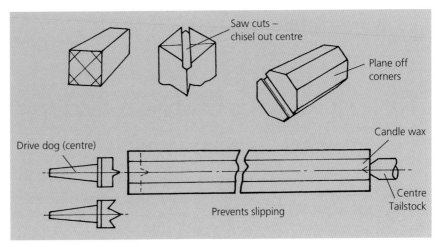

Fig. 3.61 *Preparation for turning between centres*

When turning **between centres** you need to prepare the wood so that it fixes on to the drive centre without slipping. To do this, two parallel sawcuts are made. Again, the corners need to be removed to make the initial turning easier. Also, candle wax should be applied to the tail-stock centre to help reduce friction and prevent burning (Figure 3.61).

Hints for wood turning

1 Make sure the tool rest is as close as possible to the work – remember to re-adjust it as the diameter of the work reduces.

2 Before switching on, or after any adjustment, always rotate the work by hand, to ensure no fouling or catching.

3 Select the correct speed and appropriate tool for each operation. Remember, slow speeds for initial shaping and large diameters, higher speeds for finer shaping and smaller diameters.

4 You may need internal and external calipers, as well as templates to check thicknesses and ensure accuracy.

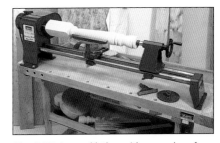

Fig. 3.62 *A wood lathe, with examples of turned work underneath*

Fig. 3.60 *Preparation for turning on a face plate*

81

REFORMING AND DEFORMING

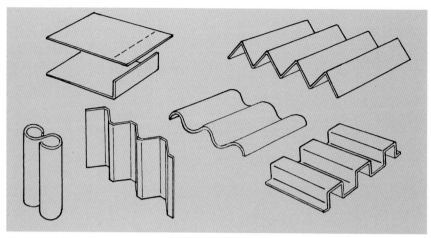

Fig. 3.63 *Folding and bending thin sheet paper and card*

If you fold and bend sheet paper or card you can produce different, stronger and more rigid structures. Changing the shape of wood, metal and plastic through reforming and deforming can be used to the same effect. **Reforming** is a process during which the state of the material changes (e.g. from solid to liquid to solid, such as casting). **Deforming** processes (such as bending, pressing, drawing and rolling) change the shape of the material without changing its state. Some materials can be worked on when cold, while others require heat to make them more workable and to utilise fully their properties (ductility, malleability and plasticity).

Getting material into the required shape in the first place is only the start of the forming process. Often, the work needs to be held and supported in that shape for a period of time, in which case clamping or the use of power presses may be necessary.

There are problems associated with **bending**. The process causes the convex layers on the outer surface to be stretched under tension, while the concave inner surface is being shortened under compression. In the middle is the neutral axis, where no change of length takes place (Figure 3.64). All the time the concave surface is trying to recover and spring back to its original length. When the stresses become severe, because it has nowhere to go, it deforms. Stretching is relatively easy to achieve, usually by softening the material by applying heat (e.g. annealing metals). This lowers the stresses upon the material and increases its **plasticity**.

Thin material deforms more easily than thick material. Smooth curves in material can only be made by eliminating or preventing distortion.

Fig. 3.67 *Kerfing*

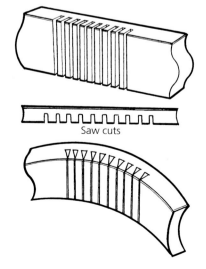

Saw cuts

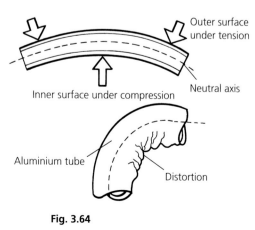

Outer surface under tension

Inner surface under compression

Neutral axis

Aluminium tube

Distortion

Fig. 3.64

Bending wood

Making curves in wood can be problematic. Simply cutting from the solid causes weaknesses, because of the **short grain** that is created (Figure 3.65). To achieve maximum strength the grain needs to follow the shape of the curve.

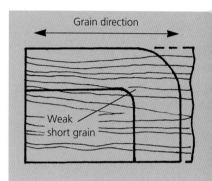

Grain direction

Weak short grain

Fig. 3.65 *Cutting a curve from solid timber creates a short grain*

Sometimes, specific pieces of wood are **selected** to meet a particular task. For example, the 'cant' (curve) of the back leg of a traditional chair occurs naturally around knots in the wood which distort the grain so that it flows around them. By making use of such areas, short grain is avoided. Unfortunately, this method is not suited to production on a large scale.

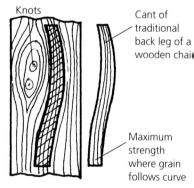

Knots

Cant of traditional back leg of a wooden chair

Maximum strength where grain follows curve

Fig. 3.66 *Selecting timber for a specific task*

Another traditional method for shaping wood, used for making curved sided musical instruments such as guitars, is **'kerfing'**. This involves saw cuts being made close together on the concave surface of the curve, which allows space for compression. Although rather basic and crude, this technique is effective and gives a smooth uninterrupted convex curve on the outer surface.

Steam bending

A bent twig, in its unseasoned 'green' state is springy because of its high moisture content. This effect can be recreated to bend solid wood by subjecting it to heat treatment – the supply of heat and moisture makes it soft, semi-plastic and compressible. A **steam chest** (Figure 3.68) treats wood placed within it with saturated steam, maintaining a temperature of 100°C. The amount of time needed for steam bending varies from timber to timber – some timbers need up to one hour per 25 mm of thickness before they can be bent. When the steam treatment is complete the work should be clamped in a jig or former as quickly as possible, while it is still pliable. It then needs to be held in its bent position to settle and dry out in a warm atmosphere. Wood that has been treated in this way can be prone to twisting.

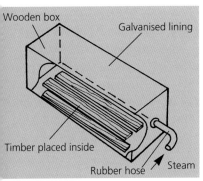

Fig. 3.68 *A steam chest*

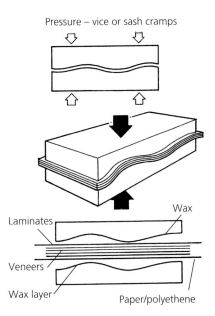

Laminating

A more accurate method of shaping wood is laminating. This involves building up the curved form with layers ('lamina'). The layers may be thin veneers, thicker constructional veneers, or saw cut strips. They are assembled so that the grain of each layer is running in the same direction, following the curve (unlike plywood, which has interlocking grain).

The layers are glued together with a **strong adhesive** (e.g. Aerolite 306) and are sandwiched between the waxed faces of a former or a jig using cramp pressure (Figure 3.69). The layers bend to the shape of the jig and 'set' together.

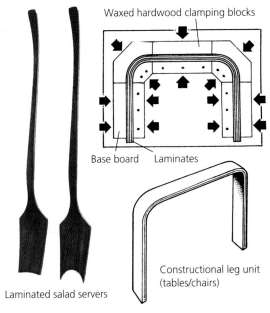

Fig. 3.69 *Laminating using clamp pressure and formers to shape veneers*

If the curve you want to make is suitable, you can use a **flexible tension band** in place of one of the formers. The band follows the shape of the former and pressure is exerted by means of a screw (Figure 3.70).

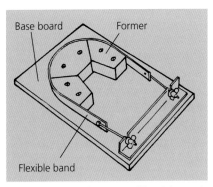

Fig. 3.70 *Using a flexible steel band for semi-circular laminating*

Another method of applying pressure is by using a **bag press** (Figure 3.71). This technique is most useful for forming small curves with thin lamina. The glued layers are taped to a former and placed on to a prepared base board, which is inserted into a bag. The bag is then sealed and the air extracted using a vacuum pump. The consequent atmospheric pressure acts on the lamina and presses them together.

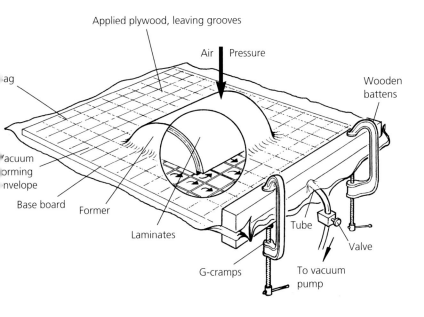

Fig. 3.71 *Using a bag press*

Richard Pullen is a jeweller and silversmith. In a small workshop behind his shop in Lincoln he works on commissioned work and repairs to antique silverware and jewellery. Most of the tools and processes used are traditional, and often centuries old.

The silver-plated chalice that can be seen being made here has been commissioned for a church. The chalice, which can be seen on the right, is made from gilding metal, an alloy of copper and zinc, that will later be silver plated. It is made up in three sections: the base, the stem (also called the 'knop') and the bowl.

In the series of photographs on the opposite page you can see the stages in the manufacture of the chalice. While it is being made, the metal needs to be annealed at regular intervals so that it remains malleable and does not split or crack.

Beating sheet metal

For any of the methods of beating metal, a **malleable material** is needed. Silver is ideal, but expensive. The most frequently used are copper, brass, nickel silver and aluminium. Constant beating causes strain on the metal and **'work hardens'** it. Therefore, it needs to be softened periodically by **annealing**. This involves heating the metal to a dull red heat (except aluminium, which requires a lower heat) and then allowing it to cool. After annealing, surface oxides form which can be problematic and leave permanent marks if hammered into the surface. To prevent this, the metal needs to be **'pickled'** in dilute sulphuric acid to remove the oxide layer. The acid acts faster when the metal is still warm, but not hot.

An alternative **'mechanical' method** of cleaning is to use an abrasive pumice powder and steel wool.

⚠ Safety

Use brass tongs to avoid contact with acid. Protect your eyes from acid splashes. Remove acid residue with warm water.

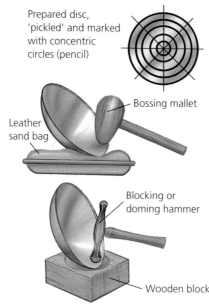

Prepared disc, 'pickled' and marked with concentric circles (pencil)

Bossing mallet

Leather sand bag

Blocking or doming hammer

Wooden block

Fig. 3.72 Hollowing

Hollowing

This process involves thinning the metal to form a shallow bowl shape. An egg-shaped boxwood **bossing mallet** is used in conjunction with a sand-filled **leather bag** or pre-shaped **hardwood block**. The blank metal disc is marked, using compasses, with concentric circles 10mm apart. These act as guides when beating. The disc is tilted on its edge and, working systematically round the disc, each circle is beaten, starting at the outside and moving towards the centre (Figure 3.72).

Sinking

Sinking is similar to hollowing, and is also used for forming shallow shapes. With this process, however, the centre portion is beaten down, leaving an unaffected flat rim or edge. A prepared **'sinking block'** is needed to position the rim and a **blocking hammer** is used to beat the metal to shape. Rim distortions will continually need to be flattened using a hardwood block and flat-faced mallet (Figure 3.73).

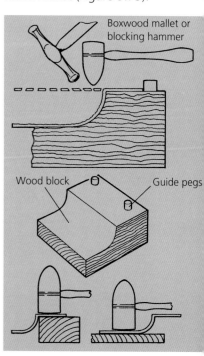

Boxwood mallet or blocking hammer

Wood block

Guide pegs

Fig. 3.73 Sinking

The chalice bowl starts out as a disk of gilding metal, cut out using tin snips and marked with concentric circles. The initial 'dishing', illustrated in the photo on the far right, takes place on a sand bag using a dome faced hammer.

After dishing the raising is carried out, using a raising stake and a raising hammer. You can see this process in the photo on the right. The raising progress is described in detail below.

On the far left, you can see the stem and base being joined together. This is achieved by soldering using silver solder with Borax as a flux.

Finally, before silver plating, the chalice is polished on a rotating polishing mop impregnated with polishing compound.

Raising

Raising is a process used to produce deep bowls and taller-sided ware. Whereas hollowing and sinking stretch the metal and make it thinner, raising the metal increases its thickness. However, a preliminary hollowing is often given before raising to the final size and shape. For shaping, a **round-headed stake** and **raising mallet** are used (Figure 3.74). To work the metal, you start at the centre and beat outwards.

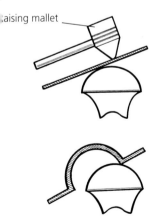

Fig. 3.74 *Raising bowls*

It is important to avoid trapping the metal between the mallet and the stake, as this will stretch the metal. To force the thickening of the metal, you need to tilt the work slightly and strike it a little forward of the point of support (i.e. slightly above

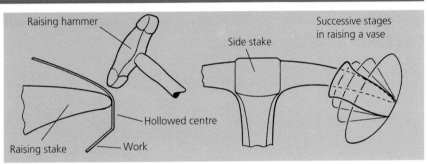

Fig. 3.75 *Raising steep-sided shapes*

the stake) so that you push the metal down on to the stake. Again, work in concentric circles. To make steeper-sided objects, such as a vase, a special **raising hammer and stake** are used (Figure 3.75).

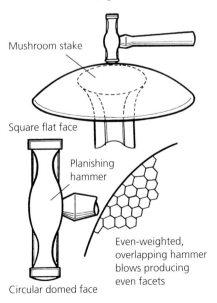

Fig. 3.76 *Planishing*

Planishing

Whatever process you use to beat metal, once you have achieved the final size and shape, you need to planish your work to remove any blemishes. This process gives the work an accurate finish, and work hardens the material, giving it mechanical strength. Planishing requires special, highly polished planishing hammers and stakes, selected to suit the profile of the shape. Following concentric circles, tap the cleaned work with light overlapping blows, trapping it each time between the stake and the slightly domed face of the planishing hammer. This will create smooth **'facets'** on the surface which can either be left as a decorative finish, or be polished out.

Fig. 3.77 *A planished silver ladle*

GKN Sankey are based at Telford where they employ 2300 people. Telford is a 'new town' that has been named after Thomas Telford, the early nineteenth-century civil engineer. It is the centre of an area that saw the start of the industrial revolution and has, therefore, a great engineering heritage.

The Engineering Products Division of GKN Sankey specialises in difficult-to-work structural fabrications made from heavy gauge sheet steel. Products such as chassis and suspension members for cars, vans and 4x4 off-road vehicles (like the Range Rover you can see on the top right). These are the parts of a car that are often not seen but that give the vehicle its strength. A high level of dimensional accuracy has to be maintained to ensure the correct alignment of all the parts of a vehicle that come together around these main structural parts.

Sheet steel components start out as flat sheet or coil strip. Often the first process is to blank out the shape required. The blank is then placed into a press where it is formed using immense force. The photograph on the right shows the blank and finished component of a Ford Transit underbody member.

Cold working sheet metal

Boxes and trays, as well as more complex shapes such as cones, cylinders and pipework can be made from sheet metal. A range of metals are suitable, including mild (galvanized) steel, aluminium, tin-plate, brass and copper. Before you go ahead and shape the metal, it is sensible to make a **card mock-up** of the developed form to determine the position and sequence of bends (Figure 3.78). At this stage you will also need to consider jointing and the size and position of any 'flaps'.

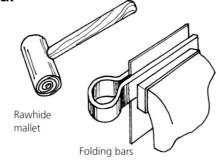

Rawhide mallet

Folding bars

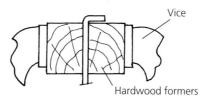

Vice

Hardwood formers

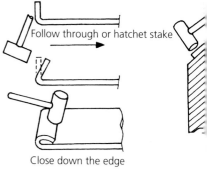

Follow through or hatchet stake

Close down the edge

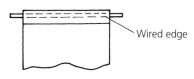

Wired edge

Fig. 3.79 Making a safe edge

To avoid exposing sharp edges, you need to make **safe edges**. This involves folding back a flap of the material, and also serves to stiffen the edge. Use **folding bars** held in a vice to bend the sheet to 90° and finish off on a flat surface, using a rawhide mallet or nylon hammer (Figure 3.79). You will need to work evenly along the edge to avoid any wrinkling. Hardwood formers can also be used to make safe edges, and for extra strength a wired edge can be made in the rolled edge.

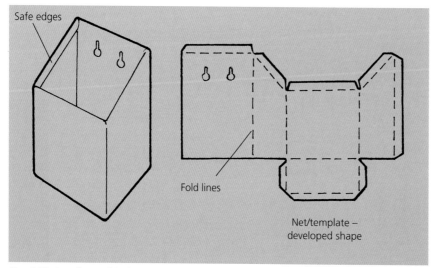

Safe edges

Fold lines

Net/template – developed shape

Fig. 3.78 A wall pamphlet holder – development and final form

On this page you can see the press and press tool that are used to complete the forming and piercing of the underbody member. This particular press forming operation requires 1000 tons to manipulate the 3.5 mm thick high strength steel. This is a complex pressing. Press tools like this are able to form, pierce and draw (stretch) the material into shape in a sequence of operations.

Heavy press work of this nature traditionally requires lubrication of the tool and the material in order to ensure that the steel moves easily across the surface of the tool without tearing. However, due to environmental concerns this process is gradually being eliminated by using carbide impregnated tool surfaces.

Press tools have to be made from hardened high-carbon steel known as 'tool steel'. They are often chrome plated to further resist wearing of the surface.

Press work and 3-dimensional forms

Pressing a product from thin sheet metal gives it a very strong shell structure. Many everyday products, ranging from pans and kettles to car bodies and aircraft panels are manufactured in this way using huge presses that exert loads of hundreds of tonnes. To form the shape, expensive precision dies (moulds) are needed. These act as a pair, with male and female halves, that allow for the gauge (thickness) of the metal.

Fig. 3.81 The manufacture of a stainless steel sink involves various pressing operations

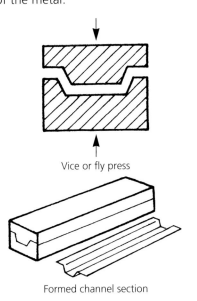

Vice or fly press

Formed channel section

Fig. 3.80 Simple two-part press tool

There is a danger of **wrinkling and tearing** when using 3-dimensional deforming, so the design limitations of the process need to be taken into consideration. For instance, sudden changes in the direction of the material (e.g. sharp bends) need to be avoided, as do **deep 'draws'** (e.g. to make a bowl shape) which can over-stretch the 'walls' of the product.

Simple two-part press tools can be made from steel (Figure 3.80). Many industrial press forming tasks combine different operations. For example, the making of a stainless steel sink involves bending the outside flanges as well as deep drawing to form the sink bowl. Further subsidiary drawing forms the ribs of the draining area, and piercing operations are needed to form the tap and outlet holes.

At GKN Sankey, transfer presses are used to carry out a series of press operations. These huge machines are fitted with a line of press tools arranged in the sequence of the operations. The component is loaded on to the press for the first operation from either a coil of steel or as a blanked shape. Blanks are loaded using pneumatically-powered handling systems, synchronised with the press operations.

Pneumatic handling is particularly appropriate for sheet material because the material can be lifted off by suction cups.

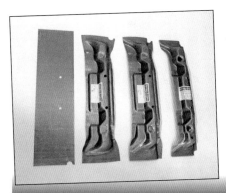

'Arms' that run along the side of the press lift the component from stage to stage as the press tool rises. The photographs on this page show a 2000 ton tri-axis transfer press and the sequence of stages that form a Ford Escort suspension unit. The part is loaded in the form of a blank, followed by two forming operations – a trim and pierce operation which produces holes for the steering components, and a flange and re-strike to enhance the strength and dimensional accuracy of the part.

Shearing

Sheet material, especially metal, is cut using tin snips or bench shears. The shearing action is like a pair of scissors, with one blade passing another, separating the waste from the work. In practice there needs to be a small gap (clearance) between the two cutting edges (Figure 3.82). The size of gap is related to the thickness of the material being cut. Generally it should be approximately 0.1t (10% of the material's thickness). The quality and ease of cut can suffer badly when this is incorrect.

Fig. 3.83 *Straight and curved tin snips*

There are several tools you will use in design & technology which remove material by shearing. The most common will probably be scissors. Sheet materials such as copper aluminium and thin mild steel sheet can be cut relatively easily using tin snips or bench shears, which are available with either straight or curved blades (the curved ones are used to cut convex curves).

Tin snips can be difficult to use at first. Make sure that the material to be cut is pushed fully back into the blades. If you try to cut just with the tip of the shears the metal will twist and become stuck between the blades.

If you find the snips too difficult to cut with you can put one handle in a vice, as shown in Figure 3.85. You can then press down hard on the other handle, making sure that your fingers are clear of the blades.

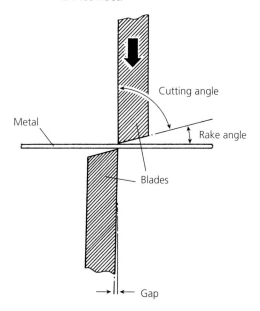

Fig. 3.82 *The shearing action*

Fig. 3.84 *Tin snips being used correctly*

Fig. 3.85

The finished components leave the press on a conveyor or a parts handling system, ready for fabrication within an assembly cell that uses robot parts-handling and computer-controlled resistance welding.

The photograph on the right shows a 2000 ton sidemember press fitted with a tri-axis transfer system – the only one of its type in Europe. It is used to press form the longitudinal members of Ford Transit vans and minibuses, an example of which can be seen in the photo on the bottom right. The material is 2 mm thick steel and the members are 5 metres long.

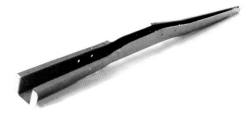

Presses are able to exert this tremendous amount of force by making use of the kinetic energy (movement energy) possessed by a huge rotating fly wheel at the top of the press. The force is directed through eccentric gears to con rods which transmit the motion to the slide.

A finished Ford Transit van can be seen in the photo on the left.

For cutting larger pieces of metal, bench shears are often easier to use as their long handle gives more leverage.

Fig. 3.86 *Bench shears in use*

⚠ Safety

Take care when using tin snips and bench shears – they can result in badly cut fingers. Also watch out for the handle of the bench shears, to make sure that it does not hit anybody standing behind the operator. Ensure that bench shears are locked when not in use and only use them with permission.

Blanking and piercing

When a number of identical components are required, blanking and piercing tools can be used to punch holes in and stamp shapes out of sheet metal. The two processes are very similar – the words used merely reflect the parts to be kept and the pieces which are waste (blanking = shapes kept, piercing = waste). Strips of metal are passed between a flat-faced hardened alloy steel **punch** and a matching **'die'** hole. The punch is forced through the strip and shears the metal on the edges of the die. The whole shape is pressed instantaneously from the strip in one movement. This requires heavy pressure for a short time, and the process is often automated. Blanking and piercing operations are often combined in a progression tool, such as that shown in Figure 3.87 and in the transfer press above.

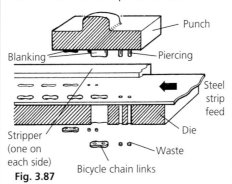

Fig. 3.87

Labels: Punch, Blanking, Piercing, Steel strip feed, Stripper (one on each side), Waste, Die, Bicycle chain links

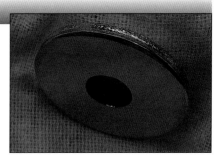

Fig. 3.88 *A washer*

If you look carefully at the edges of mass produced washers (Figure 3.88) you will see how the technique works. Two distinct textures are visible. One part is brightly cut and fairly smooth, where the metal has been sheared. The other part is duller with a rougher texture that is formed as the load increases, with the resulting deformation (tearing) from the surrounding metal.

Simple blanking and piercing tools may be used to make identical units for silver jewellery (such as chain links) and operated using an engineers' vice.

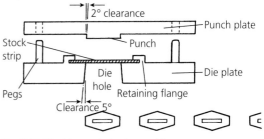

Labels: 2° clearance, Punch plate, Stock strip, Punch, Die hole, Die plate, Pegs, Retaining flange, Clearance 5°

Fig. 3.89 *Blanking and piercing tool for making identical units for jewellery*

Deforming plastics

Deforming produces a change of shape without the loss of any material, unlike wasting. Some materials, such as thermoplastics, lend themselves to deforming more easily than others. **Acrylic** is a popular thermoplastic which is readily available in sheet, tube, rod and block form. It can be transparent or translucent, and is also available in a wide spectrum of opaque colours. Acrylic can be easily deformed and reshaped at low temperatures of about 160°C. (A small fan oven, which allows air circulation and even heat distribution, or a strip heater is suitable for heating acrylic.)

⚠ Safety

Wear leather gloves when handling 'hot' plastic.

When heated, acrylic becomes pliable enough to be compressed into thin strips and even to be tied in knots! When it is reheated it endeavours to return to its original shape, which can be useful if things go wrong! Three dimensional shapes can be built up using layers of thin card – a card template is pressed into the surface of the flexible acrylic using blocks, held in a vice or press (Figure 3.90).

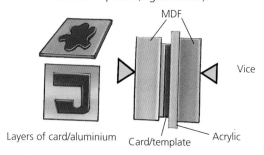

Layers of card/aluminium   Card/template   Acrylic   MDF   Vice

Fig. 3.90

Fig. 3.91 *Reverse letters ('m', 'p') on a template for a sunken pattern in opaque acrylic*

If clear transparent acrylic is used, colour can be added from the back into the depression created so that the decoration shows through to the front. Details need to be added first, followed by any background colour. Use enamel paints or permanent marker pens, which leave a smooth undisturbed front. When opaque acrylic is used to create a sunken pattern, colour can also be run into the depression. However in this case, if lettering or numbering is used, care must be taken to reverse it on the template, so that it will read correctly (Figure 3.91).

Acrylic's **'memory'** characteristic can be utilised to make a **decorative pattern** that sticks up above the surface of the plastic (this is known as 'relief') as shown in Figure 3.92.

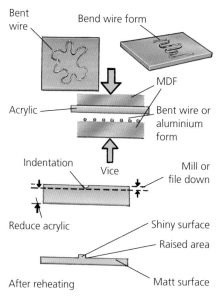

Bent wire   Bend wire form   MDF   Acrylic   Bent wire or aluminium form   Indentation   Vice   Mill or file down   Reduce acrylic   Shiny surface   Raised area   After reheating   Matt surface

Fig. 3.92 *Creating a pattern in relief*

Small diameter rod can be bent into any shape and the oven softened acrylic, together with the shaped rod, is placed between pieces of plywood or MDF and then clamped in a vice and held until it has 'frozen'. Removing the metal rod leaves an indentation of the same shape. To create a **pattern in relief** from the indentation it is necessary to file down most of the surface with a file or milling machine, until just a small indentation is left. When the plastic is reheated, the indentation

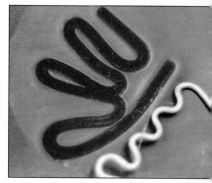

Fig. 3.93 *Raised form*

returns to its original height, so that the sunken shape is raised above the filed-down surface. The filed part aquires a matt surface, unlike the shiny surface of the original acrylic which has been raised (Figure 3.93).

Line bending

Fig. 3.94 *A strip heater*

To bend sheet plastic you need to soften it along the line that you want to fold. This involves using a **strip heater**, which heats the plastic sheet in a concentrated line with an electric element. Some strip heaters, like the one in Figure 3.94, have adjustable rests to vary the distance between the heated element and the plastic – the further from the element the wider the strip heated and therefore the gentler the bend. Do avoid making sharp bends, as this can cause thinning of the material and weaken it. Before you heat and fold the plastic, it is a good idea to make a card model/template to work out the position and sequence of folds (Figure 3.95). Also, practise heating a small waste piece of plastic first.

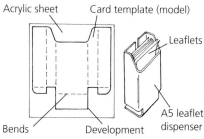

Acrylic sheet   Card template (model)   Leaflets   A5 leaflet dispenser   Bends   Development

Fig. 3.95 *A card development used as a template for a leaflet dispenser*

The position of each fold is marked on the plastic with a fibre-tipped pen. The length of time that the plastic needs to soften will depend upon its colour as well as its thickness. It is important to turn the sheet continuously so that you heat the plastic from both sides, otherwise overheating and blistering will occur. When the heated strip becomes flexible it can be bent into the required shape. Using a **former** or a **bending jig** will ensure that the angle of the bend is accurate, as well as hold the acrylic in the correct position until it 're-sets' (Figure 3.96).

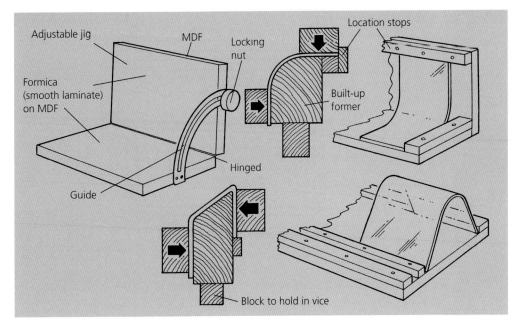

Fig. 3.96 *Formers and jigs are used to make accurate bends in acrylic*

Formers and jigs must be well prepared and have a good surface finish. Covering them with card or other smooth material will prevent wood grain and other blemishes leaving permanent impressions on the plastic. You can make formers and jigs from a range of materials – wood, MDF, piping, stakes, etc.

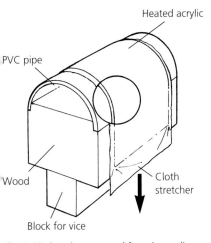

Fig. 3.97 *Creating a curved form in acrylic*

Other thin sheet plastics (i.e. polystyrene and ABS) can be heated locally with a hot air gun.

More complex curved forms, such as small trays and dishes, can be made from thin acrylic sheet using a **plug and yoke press-forming technique** (Figure 3.98). For this, a two-part former is required. The thickness of the material has to be taken into account – the gap between the male and the female part of the former needs to be

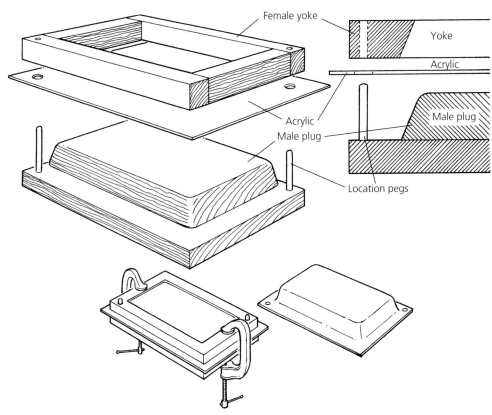

Fig. 3.98 *Deforming acrylic using a plug and yoke technique*

about one and a half times the thickness of the plastic sheet. The **male 'plug'** needs to have a slight taper with rounded corners and edges to assist removal. The **female 'yoke'** is pushed on to the plug with the softened plastic sheet in between. To make sure that the two parts of the former align correctly, dowel pegs can be fixed in the plug that fit in to holes drilled in the yoke.

The size of acrylic sheet you use needs to allow for the depth of draw (i.e. the amount that is going to be pulled down to form the sides). When the plastic has been oven heated and is pliable it is draped over the plug. The yoke is then pressed down over the plug and held there in a vice or with clamps until the acrylic cools and 'freezes' in the newly formed shape. It can then be trimmed as required.

Vacuum forming

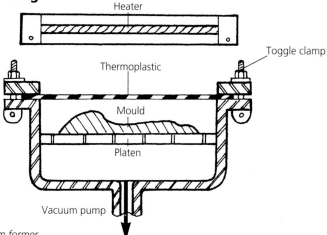

Fig. 3.99 *A vacuum former*

The vacuum forming process works by removing air from underneath soft and flexible **thermoplastic sheet** – allowing atmospheric pressure to push the plastic down on to a mould (Figure 3.99). Various packaging items, with complex deep shapes can be formed using this process, including trays, dishes and masks. Most of the common thermoplastics are suitable – polythene, PVC, high-density polystyrene, ABS and acrylic. There are different types of vacuum forming machines, which affect the size, capacity and shape of the work.

The quality of the **mould design** will determine the result – male and female moulds require slight tapering, with any sharp corners rounded, to allow the formed shape to be easily removed. The material from which the mould is made must withstand heat and slight pressure, and have smooth, mark-free surfaces to avoid showing faults. **'Venting'** (Figure 3.100) may be necessary to ensure that all the air between the mould and the material is evacuated as quickly as possible. Venting involves drilling small diameter holes, counter-bored from the underside, in awkward positions (i.e. depressions and cavities) where air might become trapped.

Thermoplastic sheet is clamped firmly in place, forming an air tight seal. Radiant heat is applied from a moveable element above the plastic sheet. When plasticity is achieved, air is evacuated from the chamber by means of a **vacuum pump**, and the resulting air pressure above the sheet forces it tightly around the mould, which rests on the platen. Deep moulds can cause problems, such as thinning on the side surfaces as the plastic is drawn down.

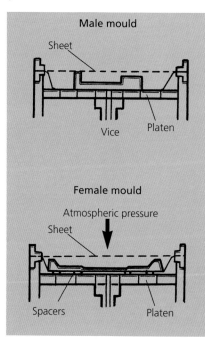

Fig. 3.101 *Male and female moulds*

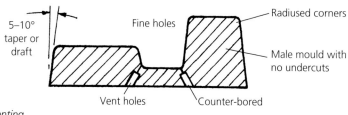

Fig. 3.100 *Venting*

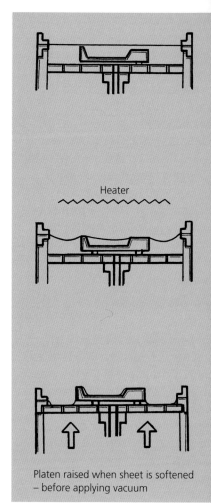

Platen raised when sheet is softened – before applying vacuum

Fig. 3.102 *Drape forming*

Drape forming helps to overcome or reduce the thinning caused by deep shapes. This process forces the mould into the softened sheet using a ram – a lever raises the platen inside the vacuum chamber. This means that the shape is partly formed (i.e. draped) before the vacuum is applied.

Fig. 3.103 *This mask was produced by vacuum forming*

Blow moulding

Similar to vacuum forming, this process involves forcing softened **thermoplastic** into contact with a mould by blowing compressed air through a narrow inlet. It is used extensively in the manufacture of hollow plastic products, such as bottles for drinks, detergents and cosmetics, and many other types of container (i.e. barrels and tanks) as well as toys.

There are many variants of the basic process. The simple freely formed dome shown in Figure 3.104 is achieved by using a clamping ring. The volume of air controls the shape, and the air pressure must be maintained until cooling takes place. A **restrictor former** (Figure 3.105) placed above the sheet can be used to modify the bowl shape. Similarities to this process can be traced to the origins of blowing glass.

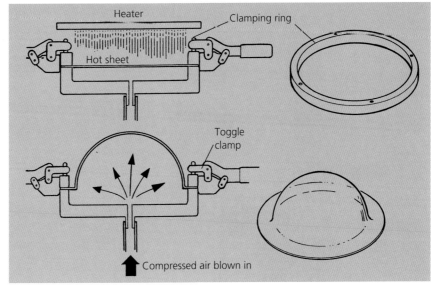

Fig. 3.104 *The blow moulding process*

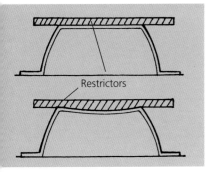

Fig. 3.105

More often, with industrial applications, the sheet is clamped over a hollow mould. The compressed air is blown on to the sheet, stretching it into the mould (Figure 3.106).

Commercial production of plastic (e.g. polyethylene bottles and containers) follows a modified process, which is illustrated on page 95. Extruded thermoplastic tube, known as a 'parison', is forced between two halves of a mould. Closing the mould seals the bottom of the tube, cutting it to the required length. Air is inflated into the soft parison, shaping it into the form of the mould. The plastic cools on contact with the mould surface and the shape stabilises. Re-opening the mould ejects the blown shape and the cycle restarts once again.

Subsequent processes include **trimming**, removing any 'flash' with a rotating blade, and printing/decoration which may take the form of in-mould printing, silk screen printing or hot-foil stamping.

Calendering

This process is used to produce plastic sheet and plastic coated fabrics and papers. A continuous flow of heated plastic is fed between a series of heated rollers which squeeze it to a consistent thickness. Finally, rollers cool and compress the sheet even more. Backing materials, such as cloth and paper can be added via another roller to make **laminates**. Patterns cut into the final rollers can also be used to imprint a design.

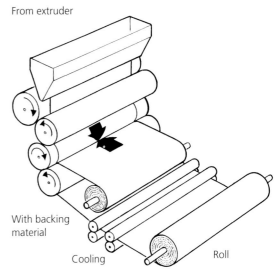

Fig. 3.107 *The calendering process*

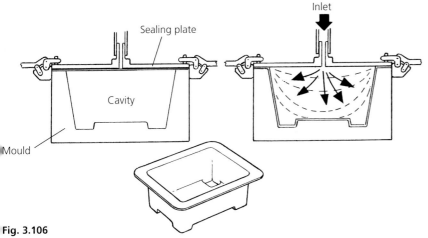

Fig. 3.106

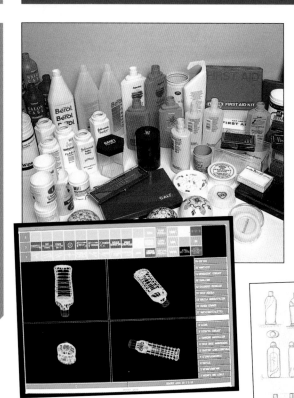

Bottle designs using line drawings and CAD software

The range of bottles that you can see in the photograph on the left have all been made using the plastic continuous extrusion blow moulding process at the Market Rasen factory site of RPC Containers. RPC Containers has many sites in England and offer a complete design and make service to customers who want plastic bottles and containers, ranging from large 40-litre polythene drums to tiny acrylic packages for jewellery.

Two plastic moulding processes are used at RPC's Market Rasen site – blow moulding and injection moulding. The blow moulding process described on the previous page can be clearly seen in the series of photographs on the next page. The product being made is a bottle for household cleaner. The material it is being made from is PVC. The sequence shows the hot molten plastic being extruded in a tubular form (top), the mould then closing around it and air being blown in via the blow pins, which forces the plastic into the mould and forms the neck area (bottom left).

Extrusion

This process is rather like squeezing toothpaste from a tube. Heated compressed plastic is forced through a die, which governs the shape of the finished product. The extruded material is then cooled to reharden it.

Plastic is much easier to extrude than metal because less force is required. Suitable **thermoplastics** for extrusion include polyethylene, PVC and polypropylene. Thermoplastics can be shaped into pipes and different cross-sections to produce insulated cable and fibres for synthetic fabrics. Irregular profiles and tubular sections are produced using extruded aluminium alloys for window frames etc. (Figure 3.110).

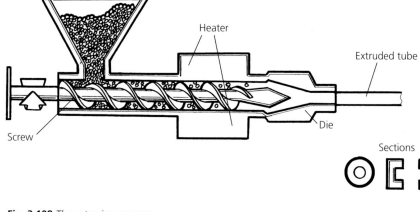

Fig. 3.108 *The extrusion process*

Fig. 3.109 *Some typical extruded products*

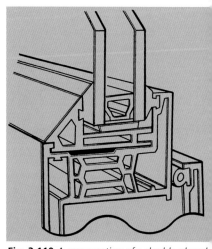

Fig. 3.110 *A cross section of a double glazed window, made using extruded aluminium*

Next, you can see waste material from the top and bottom of the bottle being cropped off (middle). In the final photo, the bottles are having air pumped into them to check that they don't leak. This is an automated inspection process, and if a bottle fails it is rejected from the conveyer belt by a pneumatic ram or an air blast. (PVC is a thermoplastic material so any rejects and cropped off waste can be recycled and used again.) The machine in the photographs produces bottles two at a time.

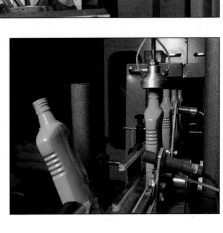

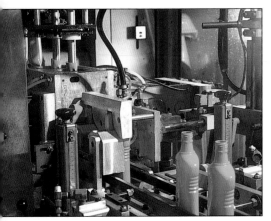

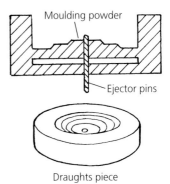

The blow moulding process in four stages: extrusion, blowing, cropping, inspection

Compression moulding

This process is used to permanently shape **thermosetting plastics** by softening and compressing them in a mould. It produces items which can resist temperature increases, such as handles and knobs.

A measured amount of powdered plastic is placed in the highly polished lower cavity of a mould, and a high pressure ram closes the mould compressing the material. The heated mould melts the material which sets hard while still in the mould. The mould is then opened and the article is removed by an ejector mechanism.

The thermosetting plastic materials used in this process include phenol formaldehyde, urea formaldehyde and melamine formaldehyde. All of these are resins in powder form.

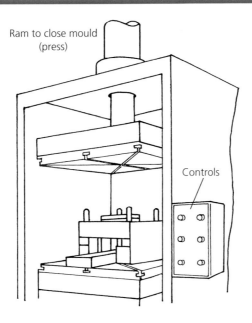

Ram to close mould (press)

Controls

Heat and pressure

Heater

Moulding powder

Ejector pins

Draughts piece

Fig. 3.111 Compression moulding machine

Car distributor cap

Hairdryer

Insulated handles

Fig. 3.112 Some typical compression moulded products

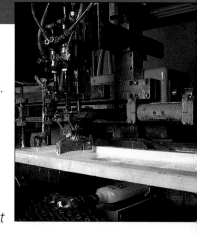

Many of the bottles made at RPC will have labels added by the customers when they fill them to show the product contained within. Other bottles, however, are printed using the silk screen printing process. A screen, which is like a stencil, is placed just above the bottle and ink is forced through the stencil on to the bottle by a squeegee moving across the screen. You can see this happening in the photo on the right. Each colour could require a different screen and an additional process, which adds to the cost of the bottle. To speed up the drying of the ink, a special ultra violet sensitive ink is used which is dried almost instantly by the ultra-violet curing process (called photo-polymerisation) which is built in to the printing machine.

Injection moulded containers

RPC produce two ranges of products using injection moulding – tamper-evident containers (such as those used for healthcare and pharmaceutical products) and containers for the promotional packaging of products like pens, coins, medals and jewellery. Injection moulding is a highly automated process whereby thermoplastic products are formed by injecting molten plastic into a mould. The process is used to produce a wide range of products including television set outer casings, buckets, flower pots and milk bottle crates.

The molten material is forced through a nozzle by hydraulic pressure that forces and squeezes the plastic into the mould where it cools. The mould is then opened to release the injection moulded product.

Plastics forming processes summary

Vacuum forming

Fig. 3.113 *Thermoplastics – polyethylene, PVC, high density polystyrene, ABS, acrylic*

Blow moulding

Fig. 3.114
Thermoplastics – polythene, PVC, acrylic

Calendering

Fig. 3.115 *Thermoplastics – PVC, Vinyl, polypropylene film, cellulose acetate*

The injection moulding machine in the photograph on the right produces tamper-evident caps used for tablet containers. The material used is polypropylene. This is because it is not brittle and has a high resistance to chemical contamination. The hoppers on the top of the machine contain raw plastic material that is mixed with a small percentage of colour granules.

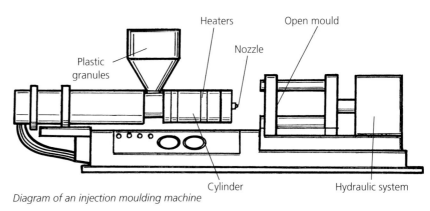

Diagram of an injection moulding machine

The tops for the containers are made on a similar machine, but they are made from polyethylene which is a little more flexible than polypropylene, and enables an airtight container to be produced. This means that the tops can be pushed on easily and the tamper-evident tag can easily be removed.

The promotional containers for commemorative coins and other items that need to be kept clean, but still be visible through the packaging, are made using clear polystyrene and clear acrylic.

People who work at RPC wear protective lab coats and hats. This is not to protect them from the machines or plastics, but to ensure that they do not contaminate the products.

Extrusion

Compression moulding

Fig. 3.118 *Thermosetting plastics – phenol formaldehyde, urea formaldehyde and melamine formaldehyde*

Fig. 3.116 *Thermoplastics – polythene, PVC, polypropylene*

Injection moulding

Fig. 3.117 *Thermoplastics – polythene, PVC, polypropylene, nylon*

GKN plc is an international group of companies employing 33,500 people in over 30 countries. The core product that they make is the 'constant velocity joint' (CVJ). This is the part of the 'driveline' assembly for cars, vans and minibuses, that transmits the power from the vehicles' engines to the wheels. They are called 'constant velocity' joints because they enable the wheels to rotate without the speed fluctuating, even when steering or bouncing about on a bumpy road.

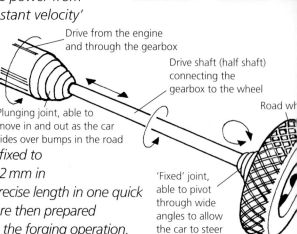

Drive from the engine and through the gearbox

Drive shaft (half shaft) connecting the gearbox to the wheel

Plunging joint, able to move in and out as the car rides over bumps in the road

Road wh

'Fixed' joint, able to pivot through wide angles to allow the car to steer

The illustration on the right shows a typical drive shaft assembly on a front wheel drive car. It consists of two joints with an interconnecting drive shaft. The starting point for the 'bell' of the fixed joint (the end that is fixed to the car wheel) is a short length of 0.53% carbon steel bar 52 mm in diameter. A hydraulic billet cropping machine shears off a precise length in one quick operation to within 1% of the weight required. The billets are then prepared for forging by being given a graphite coating that lubricates the forging operation.

HOT FORMING METAL

Forging

Forging is associated with the manufacture of strong metal components, as well as more decorative forms. The traditional highly skilled Blacksmith embraced many kinds of forging work for a variety of products – swords, ploughshares, even ornamental ironwork. In contrast, the steel forging carried out in industry today is a sophisticated, computer-controlled process.

Hot forging is one of the oldest techniques of forming metal. The process ranges from simple hand-forging, through machine drop-forging used in making hand tools, to gigantic presses for forging industrial parts. Hammering hot metal into shape improves the structure by **refining the grain** of the material, making it more dense. Strength is increased because the grain flow follows the shape of the component (see Figure 3.119). This is important, especially for strong steel components like crankshafts and gear blanks.

Force

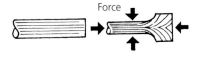

Fig. 3.119 *Hammering hot metal refines its grain*

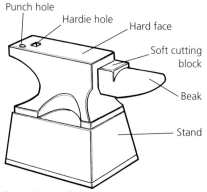

Punch hole

Hardie hole

Hard face

Soft cutting block

Beak

Stand

Fig. 3.120 *Anvil*

The techniques for hand working hot metals have changed little over the centuries – the modern **chip forge** provides the heat source to soften the metal. **Heat** is critical in all forging. It not only has to be sufficient, but for deformation to be effective, it also needs to be applied in the correct place. The spread of heat can be controlled by quenching some areas with water.

The main workspace for forging is provided by the **anvil** (Figure 3.120) which has areas suited to different operations. The face is hardened (high carbon steel) and there is a softer cutting block. The hardie hole locates and supports other tools.

Forging processes include **'drawing down'**, which increases length by reducing the section. It is illustrated by: (a) **using fullers** (Figure 3.121),

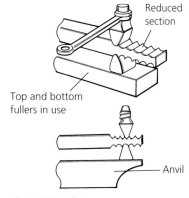

Reduced section

Top and bottom fullers in use

Anvil

Fig. 3.121 *Fullering*

(b) **flattening**, by smoothing rough surfaces (Figure 3.122) and (c) **swaging**, by forming circular sections (Figure 3.123).

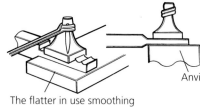

The flatter in use smoothing a rough surface

Anvil

Fig. 3.122 *Flattening*

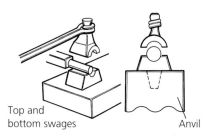

Top and bottom swages

Anvil

Fig. 3.123 *Swaging*

The forging process is a 'warm form process'. Although it is called 'warm' forming it is carried out at 920°C so the steel is, in fact, red hot. Hot forming processes are those that occur at 1200°C, at which temperature scale is formed on the surface and structural changes take place within the metal so it is an advantage to work the steel 'warm'.

The billets are heated using an induction furnace. Induction heating is very common within manufacturing industries because it is quick and very efficient. The billets pass through a high frequency, high amperage (1000A) electric coil that 'induces' heat within the metal through excitation of its molecular structure. This is a continuous conveyer process that then feeds the billets directly into 1600 ton presses that forge the steel into shape in four quick operations.

The tools used by the press in the forging process are called punches and dies. The punch forms the inside shape by pressing the billet into the die which at the same time forms the outside shape.

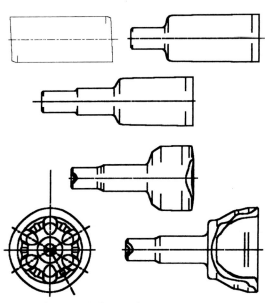

The stages in forging the CVJ outer race

Upsetting

'Upsetting' ('jumping-up') increases the section by reducing length.

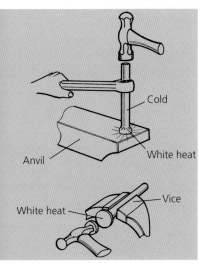

Fig. 3.124 *Upsetting*

Allowances can be made in this process for sharp bends without losing any thickness (Figure 3.125) or for decorative features.

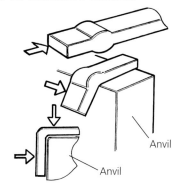

Fig. 3.125

Bending

Another forging process, bending, is illustrated in Figure 3.126, which shows the stages involved in forming a loop.

Making holes

Drilled holes weaken metal because they cut the flow of the grain. Punched holes, where the grain flow is diverted rather than cut, reduce strength far less (Figure 3.127). Punching is an important technique used at the ends of structural ties. 'Drifting' is a process used to enlarge and re-shape punched holes.

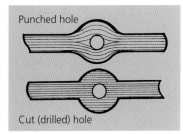

Punched hole

Cut (drilled) hole

Fig 3.127

Twisting and scrolling

Twists and scrolls are traditional decorative features used in wrought ironwork. The technique (Figure 3.128) requires skilful hand–eye co-ordination.

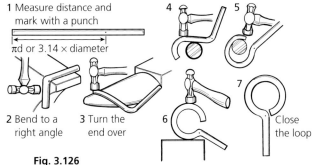

1 Measure distance and mark with a punch

πd or 3.14 × diameter

2 Bend to a right angle

3 Turn the end over

4

5

6

7 Close the loop

Fig. 3.126

Twisting

Twisting wrench

Vice

Hot

Scrolling

1 Draw down — Red heat

2 Flatten and taper

3 Start scroll by turning over the tip

4 Continue by rolling-up the face

5 Vice — Use horns to form the scroll

6 Scroll tool — Vice — A scroll tool can be used to complete the scroll

Fig. 3.128 *Twisting and scrolling*

At GKN Automotive, the machining processes that turn the forged outer race into the finished component ready for assembly are carried out on a high-volume automated flow line using robotic parts-handling and CNC-controlled machine tools. First, the forged outer races are loaded onto individual magnetically coded carriers. The code on the carrier enables the progress of each individual component to be traced throughout the system. The carriers move on a track around

Red hot CVJ outer races

a continuous circuit, rather like a model railway. Points are set in the circuit track that automatically stop the carrier when it arrives at the next process required, and move it into a short siding to await the robotic arm that will load the component on to the machine for processing, after which it is returned to its carrier.

Punch and die set used to forge the CVJ outer race

The centre lathe

The centre lathe is used to make cylindrical components in metal and plastics. Lathes are amongst the earliest forms of machine tools. The basic principles, first developed in the nineteenth century, remain in use today. Material is held firmly and rotated while a cutting tool, supported in a tool post, cuts using the simple wedge cutting action. The final shape of the material depends upon the path taken by the tool. Nowadays, however, these processes can be controlled using CNC equipment.

It is very important to **hold work** firmly and securely, as the workpiece may be rotated at high speed as well as subjected to large cutting forces.

Fig. 3.130 *4-jaw chuck*

The **self-centering 3-jaw chuck** (Figure 3.129) is the most common device used to hold cylindrical or hexagonal work. The jaws are stepped to enable bored work to be held. A second set of jaws stepped in the opposite direction further

increases its versatility. Turning the chuck key operates an archimedes spiral, which closes the jaws simultaneously. An **independent 4-jaw chuck** (Figure 3.130) is designed to hold square or rectangular sections, as well as irregular shapes. Each jaw is adjusted independently, and it can be time consuming to set up.

To hold more awkward shapes it may be necessary to clamp them to a **face-plate**, using angle brackets. Counter weights will need to be added to maintain balance.

Fig. 3.131 *A face-plate, used for mounting irregular shapes*

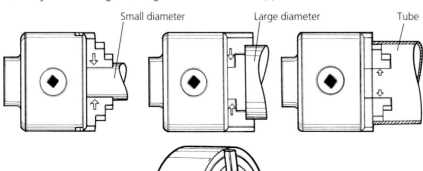

Small diameter　　Large diameter　　Tube

Detail

Fig. 3.129 *3-jaw chuck*

The first process along the line is to centre drill the thin end of the outer race on a small automatic lathe. It is then loaded onto a multi-tooled CNC lathe, held from the inside of the bell shape and supported at the other end by a centre. In all of the processes component chucking is pneumatic.

The machining process uses tungsten carbide tipped tools that are held in an indexing tool turret that rotates the required tool as appropriate. The whole process of machining all of the outside form takes less than one minute. Each flow line has two CNC lathes performing this operation.

A robot off-loading forged outer races from pallets

After turning, the races are subjected to statistical process control (SPC). This involves the automatic checking of the size of every third component to see if there is a trend towards moving out of tolerance. It indicates when a tool tip is wearing out or damaged, and so needs changing (see page 27).

A CNC lathe turning the outside of the outer race (this process is normally guarded and awash with coolant)

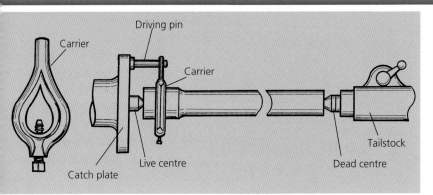

Fig. 3.132 *Turning between centres*

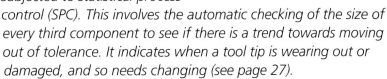

Fig. 3.134 *Turning tools*

Long pieces need to be held **between centres** (see Figure 3.132). The material must be prepared by first facing and then centre drilling both ends to accommodate the centres. A **'live' centre** is located in the spindle headstock. An appropriate sized **carrier** ('driving dog') is clamped to the workpiece. The drive is transmitted from a catch plate, which is screwed on to the spindle, to the carrier. Ideally, a revolving centre with an internal ball race supports the other end in the tailstock.

Turning tools used on a centre lathe (Figure 3.134) need to be both hard and tough. High carbon, high-speed steel, or even more resistant carbide tipped tools are used. The shape of the tool used depends upon the operation to be carried out. A **roughing tool**, preferably radiused,

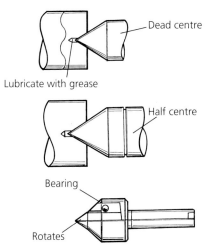

Fig. 3.133 *Centres*

can reduce the material to within 1mm of its required finished size. **Fine finishing cuts** are achieved using a small radiused point. **Knife tools**, left or right, enable the cutting of sharp corners. A **parting tool** is used for making grooves and cutting off the work from the material remaining in the chuck.

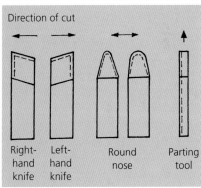

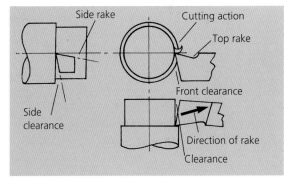

Fig. 3.135 *Setting the tool correctly*

To ensure the **efficient cutting** of all material, the tool geometry must be suitable for that material. This means that the **clearance angles** (the angles that enable only the cutting tip to be in contact with the material) and the **rake angle** of the tool (which is the top angle, measured along the line of cut) must be correct. The latter also requires correct tool setting with the tool point accurately located on centre height.

The process of forming the spline and the screw thread on the stem end of the outer race is called 'rack rolling', illustrated in the photo on the left. Within high-volume manufacturing, screw threads are rolled rather than being cut. This is a much faster process and the grain flow of the metal is improved. It involves squeezing screw thread into the end of the shaft between either rotating rollers or, as in this case, racks. The top and bottom rack contains the thread form and the form of the spline. They move in opposite directions spinning the stem end of the CVJ outer race and pressing the form into it. This is a very quick process in which no metal is actually removed because the shape is squeezed in and out of the metal.

The photo on the left shows a CVJ outer race after rack and rolling – the splines are for the location of the wheel assembly and the screw thread is for the wheel centre nut.

Turning speeds

A further critical feature is the turning speed (the speed the lathe should rotate). This is determined by the material and the diameter of the work. The formula in Figure 3.136 and Figure 3.137 provide a useful guide for cutting speeds.

$$N = \frac{1000\,s}{\pi d}$$

where N = Speed in revs per minute
s = Cutting speed (metres per minute)
d = Diameter of work

Example – the turning speed for a Ø40 mm mild steel bar

$$N = \frac{1000 \times 25}{\pi \times 40} = 200 \text{ rpm}$$

Fig. 3.136 *Cutting-speed formula*

Most machining processes benefit from using a **cutting lubricant**. This acts as a coolant, reducing friction and heat and helping with waste removal. Recommended lubricants are shown in Figure 3.137.

In addition to basic turning, more sophisticated processes such as taper turning and screw cutting can be undertaken on some lathes. Another simple and useful process is

Material	Rake angle	Cutting speed M/min with HSS tool	Cutting lubricant
Aluminium	40°	200	Paraffin
Brass	2°	90	None needed
Cast iron	2°	20	None needed
Hard steel	6°	18	Soluble oil
Mild steel	20°	25	Soluble oil
Nylon	30°	170	None needed
Acrylic	40°	200	Paraffin

Fig. 3.137

'knurling'. This involves making a pattern in the surface of cylindrical material, usually to provide a textured grip. A specialist tool with hardened wheels is fitted into the tool holder and then pressed into the rotating material. It can then be moved across along the surface.

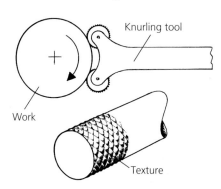

Fig. 3.138 *Using a knurling tool*

Accurate drilling can also be undertaken using a centre lathe. It involves the workpiece rotating while the drill remains stationary. The rotational speed of the material is determined by the size of the drill rather than the size of the work. The drill is held in a **Jacob's chuck**, which has its morse taper located in the tailstock. A morse-tapered shank drill can also be used, but a centre drill is always necessary to start any drilling operation on the lathe. Boring, using special boring bars, can also be used to enlarge drilled holes.

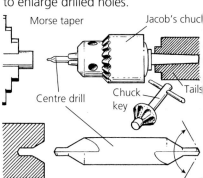

Fig. 3.139 *Jacob's chuck and centre drill*

> ⚠ **Safety**
>
> This process requires great care, as does any operation using a centre lathe. Permission and supervision, use of eye protection, chuck and chip guards is essential!

The next flow line process is the induction hardening of the splines and bearing surfaces. (The diagram on the right explains the induction heating process.) A hard surface increases strength and is better able to withstand wear when in use. The outer races are hardenable because there is sufficient carbon within the steel (0.53%). The process involves raising the temperature to red hot (in excess of 900°C) over the required area, and then quenching it in water. This is all carried out in a continuous process that lasts just a few seconds. It is a great advantage of this process that it enables only selective areas of the surface to be hardened. It is also cost effective, and leaves other parts, such as the screw thread, tough.

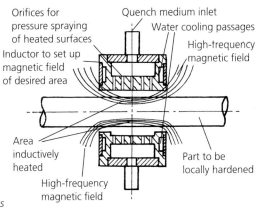

Section through the surface induction hardening process

After the outer race has been hardened, the final machining process is the precision grinding of the locating faces and the ball bearing tracks. Surface grinding is the only machining process option, now that the surfaces have been hardened. The process is very precisely controlled in order to ensure the necessary accuracy of size and quality of finish.

Precision surface grinding

Milling

This process, which can be computer controlled, uses rotating multi-toothed cutters to shape a range of materials, especially metals and plastics. Milling machines are robust and powerful. There are two main types – **horizontal** and **vertical**, so-called because of the milling cutters' axis of rotation. Horizontal machines have their cutters mounted on a horizontal arbor, which is supported within the body of the machine at one end and by a support bracket on an overarm at the other. Vertical machines have cutters mounted in specially designed vertical chucks. The lower half of the machine consists of a **machine table**, with 'Tee' slots, enabling a machine vice to be bolted down. Alternatively, work can be clamped directly to the table using nuts, bolts and clamps. The machine table can move and be locked in all three axes. The use of powered traverses in some

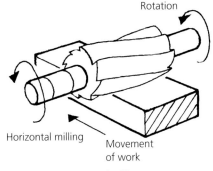

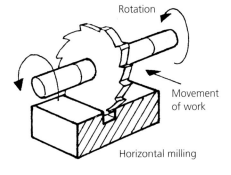

Fig. 3.141 *Horizontal milling cutters*

machines is restricted to the X axis (i.e. along the length of the table).

The **cutting action** is achieved by multi-toothed cutters, with each tooth in turn cutting a chip. **Side and face cutters** cut on both the side and the diameter.

Slots can be made using either type of machine. Vertical machines use **end mills**, but they are unsuitable for cutting down into the work unless a hole is drilled first. Alternatively, a **slot drill** is used, which is similar to end mill, but can also cut downwards like a drill.

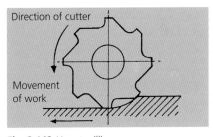

Fig. 3.142 *Upcut milling*

Cutter rotation and the direction of movement of the workpiece is important. With conventional (upcut) milling, the movement of the work is in the opposite direction to the rotation of the cutter. This produces a wedge shaped chip that starts fine and becomes thicker.

Fig. 3.140 *Vertical milling*

> ⚠ **Safety**
>
> In the illustrations, guards are removed for clarity purposes only. Never operate without permission. All work must be clamped securely to the machine table. Make sure all guards are in place and eye protection is always worn.

In 1913 John Rundle bought a steam driven tractor and a trailer, and became a haulier. Later, he became a contract thresher travelling around local farms threshing their grain. His business expanded and after the Second World War the company went into agricultural engineering. In 1950 they opened a foundry on the site of an old village public house 'The Globe'. It therefore became known as the 'Globe Foundry'.

Most of their products, including agricultural equipment, reproduction Victorian seats, ornaments, and brackets and bollards for sea side railings, are made from cast iron, although some brass, bronze and aluminium casting is also carried out. They are also able to repair, replace and match up any item – 'we can take a sample and produce whatever you need, that's the beauty of foundry work, it is adaptable and flexible', says Ken Rundle.

All of the raw material used for cast iron work is recycled scrap metal

CASTING

The casting process is different from other forming techniques because the material starts in a liquid state. Like jelly, it can then be poured into some kind of mould, retaining the shape and form of the mould once it solidifies.

Wasting from a solid piece to make an engine or gearbox case, for example, would mean more material being discarded as waste, than would remain in the finished product. Such a method would not only be time consuming and expensive, it would also not be very strong. By contrast, casting enables shapes ranging from the very simple to the most intricate, to be produced in forms that could not be achieved in any other way.

The process is suitable for a range of materials, including plastics and concrete, but is especially effective for metals. When molten metal solidifies, small crystal grains form in a definite arrangement which flow round the shape of the mould. Rounded edges and corners formed in casting help considerably to enhance the metal's strength. A variety of metals with different properties and melting points are used to make functional as well as

Fig. 3.143 *These planes have been made using the casting process*

decorative items. Cast iron is used for drain gratings, engine parts and tools. Precious metals are also used to make jewellery and other decorative items.

Casting using aluminium

Casting aluminium in sand moulds is the process most commonly used in school workshops. The aluminium alloy (e.g. LM4) is reasonably inexpensive and readily available. The casting temperature (up to 750°C) is not too difficult to achieve. The process involves four stages – pattern making, mould making, melting and pouring, and fettling.

Fig. 3.144 *A decorative wall plaque cast in aluminium*

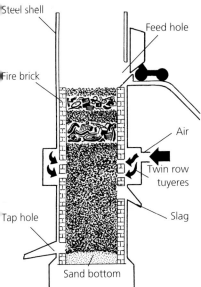

The metal is melted in a coke-fired blast furnace that is loaded as shown in the photo on the opposite page.

Layers of scrap cast iron and coke are supplemented with small amounts of silicon and manganese, which are added to replace that lost in the recycling.

Sand moulds are made using patterns made from wood or from other castings. A shrinkage of 1 mm in 100 mm has to be allowed, this means that the pattern for a 1 m diameter wheel must be 10 mm larger than the size required.

The main differences between casting in school and casting on this scale is the temperature and the scale of the operation. The cast iron melts at 1500°C and has to be raised to 1650°C to ensure a good uniform flow.

'Tapping' the furnace

Diagram of a blast furnace

Pattern making

The quality of the casting is determined by the pattern. This is made slightly larger than the required finished size to allow for shrinkage as the metal cools. There are specific rules of contraction that determine this. With aluminium, it can be up to 5 mm per 300 mm of pattern. Jelutong, an easily worked, close grained hardwood, is well suited to pattern making.

Clean and easy removal from the sand is an important consideration. Casting **'draft'** (gradually tapering the vertical surfaces) assists the withdrawal. Sharp corners need to be rounded and internal corners radiused with fillets to prevent cracking, before the final pattern is sealed with varnish or paint.

A simple form of pattern is the flat back type, where all the detail appears on one side, such as in the fish wall plaque in Figure 3.144. Cylindrical forms, like handles require a **split pattern** to avoid undercutting. They are made in two halves and located together by pegs (Figure 3.146).

Mould making

Moulding sand needs to be sieved to get rid of lumps and foreign matter. The sand must be damp enough ('green sand') to hold its shape, but not to stick to your hands. Commercially available oil-based products (e.g. Petrobond) are better than water for mixing, as problems caused by steam are eliminated. Hot metal in contact with sand chills instantly, preventing it from being washed away.

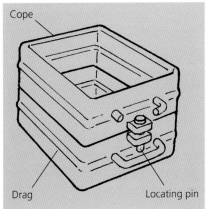

Fig. 3.147 *A mould box*

The **mould box** is a two-part metal box containing the sand. The upper half is called the 'cope' and the lower half the 'drag'. The halves are located by two pegs in the cope. The stages involved in making a mould box are illustrated in Figures 3.148 to 3.153 (overleaf).

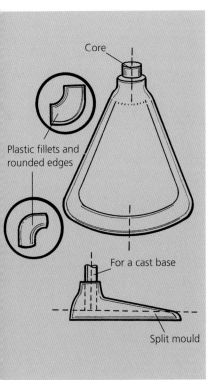

Fig. 3.145 *A wooden pattern*

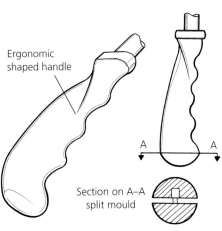

Fig. 3.146 *A split pattern for a handle*

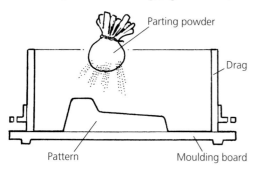

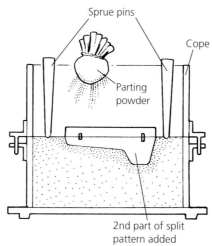

The photographs on this page show the mould and the pattern for a crown wheel for a fairground ride and the molten cast iron being poured into the mould.

Polystyrene is also used in a 'lost pattern' type of process. In this process the polystyrene pattern is buried in the sand and is then completely burnt away by the incoming melt. The sequence of photographs on the next page shows a horses head for a gate post being made by this method. The polystyrene is buried in the sand, the melt is poured in and burns off the polystyrene, and the casting is then removed.

First of all, the drag is placed upside down on a flat moulding board. The half pattern, without the pegs, is placed flat down on the board, at least 50 mm from the sides and sprinkled with parting powder, French chalk or a proprietary dressing, to prevent it sticking (Figure 3.148).

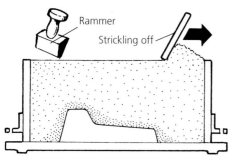

Fig. 3.148

Fine sand is sieved over the pattern until it is covered with a 25 mm layer. This is compacted by gentle ramming. More sand is then added and rammed until the drag is filled. The surplus is 'strickled off' using a straight edge (Figure 3.149).

Fig. 3.149

The drag is then turned the right way up and the cope is placed on top. The upper part of the pattern is located in position and the area, including the joint line, is dusted with parting powder. At this point, **sprue pins** (tapered circular hardwood pegs) are added. They create the **runner**, down which the metal flows into the mould and the **riser**, up which air escapes and the excess melt rises, when the mould is filled (Figure 3.150).

Fig. 3.150

Following the same procedure as the drag, the cope is then rammed full and strickled. Before removing the sprue pins the top holes are cut into a funnel shape and smoothly rounded to provide a pouring basin (Figure 3.151).

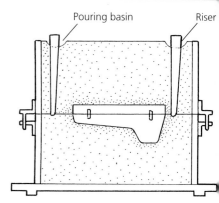

Fig. 3.151

Next, the two halves are carefully separated. Before each half of the pattern is removed, **gates** are cut in the lower half of the mould, to link the mould to the runner and riser, allowing the melt to flow without turbulence (Figure 3.152). **Vent holes** are pricked to permit gases to disperse from areas where they may become trapped.

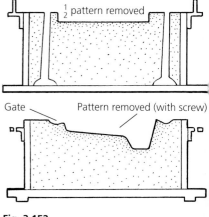

Fig. 3.152

In more complex castings where holes or cavities need to be formed, **cores** are used. These are made of sand mixed with linseed oil and baked. They are positioned at this stage and any loose sand blown clear. After careful re-assembly, the mould box can be positioned ready for pouring.

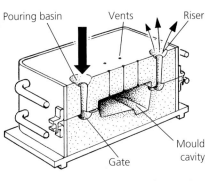

Fig. 3.153 *The mould box, ready for pouring*

Melting and pouring

Aluminium melts at 600°C, but needs to be at 750°C for pouring. Pour slowly, but continuously into the pouring basin, until a pool forms at the top of the riser.

⚠ Safety

Great care is needed! Never pour without help and supervision. Wear protective clothing – leather gloves, apron, leggings and a full-face mask and ensure good ventilation.

Fettling

When completely solidified, the sand can be carefully broken away and the casting 'fettled'. This involves sawing off the runners and risers and filing off the gates and any **flashes** formed by leakage at the joint line.

Lost pattern casting

This technique is also suited to aluminium casting. The disadvantage of this method, however, is that the pattern is not reuseable. **Expanded polystyrene**, commonly used as a packaging material for electrical goods, can be shaped using a hot wire cutter and

⚠ Safety

Ventilation is essential, as toxic fumes are given off!

built up into a pattern using PVA glue. The structure of polystyrene is made up of small soft compacted pellets, within a large volume of air. When molten metal is poured on to it, vaporisation occurs, leaving little residue. However, its own characteristic texture is left on the surface of the solidified metal. A variety of shapes and complex sculptural forms are made possible because the pattern is not removed, but left embedded in the sand.

Fig. 3.154 *A trivet, made using lost pattern casting with a polystyrene pattern*

Above the casting floor at the Globe Foundry, a ladle, suspended from a crane holds 6 cwt of molten cast iron. Pouring usually takes place twice a week, when the whole of the casting floor has been prepared. The floor is one huge sand pit that is 'worked up' after use using a garden rotivator. The weights that are placed on the sand moulds are to prevent expansion gases and moisture from pushing the top of the mould upwards and causing distortion. It takes two days to prepare the floor for casting and around two hours to carry it out. The Globe Foundry produces between 200 and 300 castings per week and nothing is wasted – spillages, runners, risers and bad castings are all recycled.

The casting floor just after pouring

Lost wax casting

This is one of the oldest techniques of casting. As the name suggests, the pattern is made from wax, which can be of unlimited shape. Various organic waxes, including beeswax can be used together with proprietary waxes that have good modelling properties (i.e. carving, shaping and dripping). A fundamental requirement of any wax is that it must melt out and burn away without leaving any residue. When completed, the wax pattern is mounted on sprues. These smooth rods of round wax leave a cavity in the plaster for the molten metal to flow into the mould. In general, sprues must be kept as short as possible but their size, number and position is determined by the pattern. The pattern is surrounded by special refractory plaster (known as investment plaster) which is able to withstand 750°C (Figure 3.155).

The pattern is removed ('lost') by heating up the mould in a muffle furnace or a burnout oven. Initially, the wax melts and then burns away. While the mould is still hot (400–700°C) molten metal is poured in under pressure. This may be in the form of air pressure, vacuum, centrifugal force, or steam pressure.

The use of this process enables very fine detail to be obtained and is usually associated with jewellery design. Intricate shapes and small sections, including filagree work, leaves, animals, insects and rings can be cast as one-offs in precious metals, such as nickel, silver, gold and platinum. These techniques are also valuable to engineers and meet the need for accuracy demanded in the electronics field.

Fig. 3.156 *A necklace, made using lost wax casting*

Cuttlefish casting

Another method used by jewellery designers to cast small detailed pieces is achieved by using a cuttlefish mould. This natural material withstands high temperatures and is easily shaped by engraving or by pressing a hard pattern into it. The texture of the cuttlefish itself is often used to advantage on the surface of the casting.

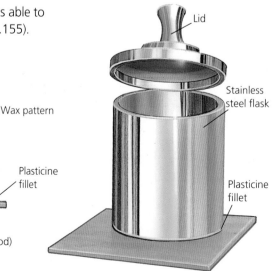

Investment plaster

Wax pattern

Sprues

Plasticine fillet

Ø1.5 wire (welding rod)

Lid

Stainless steel flask

Plasticine fillet

Fig. 3.155

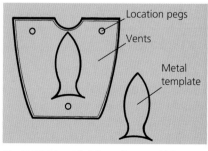

Location pegs

Vents

Metal template

Fig. 3.157 *A cuttlefish casting mould for a fish pendant*

Rundles are the only company in England who still manufacture the traditional 'up and down horses' fairground ride, known properly as 'gallopers'. They are still the same traditional attraction, the main difference being that the horses are now made out of GRP (see page 112) rather than wood. On the right you can see an example of the gallopers made at the foundry.

Cast iron products, from the Globe Foundry including a crown wheel and pinion for a set of 'gallopers'

A flat joint is constructed between two pieces of cuttlefish, which are accurately aligned by small pegs, or ball bearings. The design is cut or pressed into one or both surfaces of the mould. A funnel shaped pourer is cut to the mould cavity, and air vents are cut, radiating from the cavity, to near the outside of the mould (Figure 3.157). The cavity faces are coated with a strong solution of borax and left to dry. They can be further toughened by a solution of sodium silicate (water glass), which protects against the effects of hot metal. The mould is then wired together and a charcoal block can be added on to form a crucible for the gravity pouring (Figure 3.158).

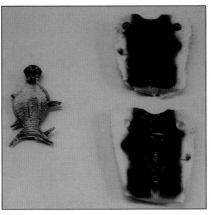

Fig. 3.159 The fish pendant, just out of its mould and final product

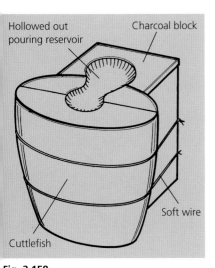

Fig. 3.158

This technique is ideal for making 'one-offs', or a very small run, using precious metals, especially silver.

Die casting

In industry, large numbers of quality castings need to be made, and therefore the moulds ('dies') used need to be permanent. These special alloy steel moulds are often made in sections to facilitate the components' removal. Because of the expense involved they are most suited to production in quantity and where accuracy of size, shape and surface finish is essential.

When molten metal is poured into the cavity under its own weight, it is known as **'gravity'** or **'permanent' casting**. If force with a ram is needed, it is called **'pressure die casting'**. Very fine detail and a high quality finish are achieved by this process. Little additional finishing is required other than removing any small 'flashes' caused by

leakage between the parts of the die. Lead, zinc, aluminium and brass alloys are cast in this way to produce toy cars, military models, camera bodies, car parts and components for domestic appliances.

Fig. 3.160 Die cast products

*Jolly Roger Amusement Rides are a company that produce coin-operated rides for young children, like the one shown in the photo below. It consists of just 23 people, and it's business stretches across the world – from Europe to Japan, Bahrain, Nigeria and North America. The products are all made from **GRP** and have a fabricated steel underframe, which supports the GRP moulding and provides the mechanical movement of the ride (usually rocking or spinning). The GRP process Jolly Roger use is hand **lamination**. There are other GRP processes that are used in industry that Jolly Roger have considered.*

The spray deposit process involves spraying a mixture of glass fibre strands and resin through a hose into a mould. This process is fast, but can be very messy and gives poor coverage on corners.

The saturated roller process entails rolling glass fibre matting on to a mould using a large roller that has been saturated with resin. This process is best suited to the production of large items, such as bus fronts and caravan roofs.

Casting resin

Polyester resin is a **thermosetting** material. The change of state, from a liquid to a solid hard mass, is known as **'polymerisation'** and takes place when the resin is treated with special chemical agents. Various formulations, including activators, are available to meet most requirements.

Most resins are sold in a pre-activated condition (i.e. the activator has already been added). All that is needed to initiate polymerisation is the catalyst. An intermediate state in the process is known as the **'gel-state'**, where the resin assumes a jelly-like state and is quite flexible and tough.

Embedding encapsulates and preserves small specimens and decorative items (coins, insects, electronic and mechanical components). A simple casting process can be used to make personal novelties such as paper weights. A variety of moulds (glass, ceramic, plastic and metal) can be used. Ease of removal is crucial. A smooth cup, or other container shape fixed to a piece of glass and sealed with plasticine is ideal (Figure 3.162). The surfaces of the mould should be highly polished with a release wax. A chemical release agent can also be used.

The working temperature for casting resin is around 20°C (68°F). Pre-activated, clear **embedding resin** (EM306PA), to which 2% by weight of liquid catalyst is added, is poured into the mould. After a short time to 'gel', objects are arranged on this layer and further layers are added. A coloured pigment (translucent or opaque) can be added to the final layer to enhance appearance. Polymerisation is accompanied by heat being given off (an exothermic reaction). The temperature rises rapidly after the resin has set, and this accelerates the curing.

Fig. 3.161 *A cast resin paperweight*

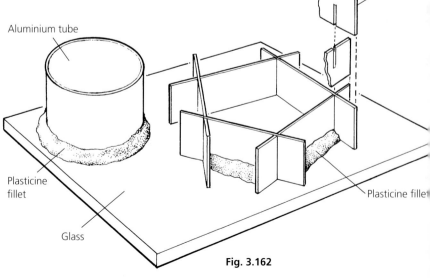

Strips

Aluminium tube

Plasticine fillet

Glass

Plasticine fillet

Fig. 3.162

Vacuum forming (in fact a cross between vacuum forming and hand laminating) involves the GRP being laid up and forced into a mould using a vacuum, which draws the resin through. This process is more suited to mass, repetitive production runs and is used for such products as motorcycle fairings.

Regular rides are designed to make the maximum effective use of coloured areas of GRP gel coat, and only have small areas of painted colour or transfers added to them. The train in the photograph on the previous page is made from yellow and black gel coat and has had transfers added to its wheels and body. Jolly Roger also make character rides based on well-known children's characters such as the Pink Panther, Mr Blobby and the Mr Men. Because they are based on licensed characters ('licensed' means that their image can only be reproduced with permission from the people who own the copyright) these rides have to be made very accurately to faithfully reproduce the character, otherwise permission will not be given. The character rides are made using multiple part-moulds, and the sections are fixed together using GRP. Unlike the regular rides, only clear gel coat is used, and the characters are finished off by spray painting and air brushing, and finally, a coat of clear gloss lacquer is applied. The photo on the right shows a pattern being made for a Mr Men character ride in the shape of a shoe by spreading car-body filler over a structure of polyurethene foam and chipboard.

Fig. 3.163

Casting can employ a variety of techniques. **Latex moulds**, such as those designed for the chessmen in Figure 3.163 do not require a release agent. The steps in the process used to make the chess pieces are given below:

1 Prepare pieces of card with a central hole to act as a support for suspending the moulds when filled (Figure 3.164).

2 **Casting resin** (FC702PA) can be coloured with pigment paste. Up to 25% of a toughening agent, 'Trylotuf' can also be added to the resin. Add a **liquid catalyst** 2% and stir.

3 Half fill the mould with the catalysed resin, squeeze between the fingers to **force out any air bubbles**, before topping up to the shoulder. Repeat the air removal process.

4 Place the filled mould into its support, and if necessary, top up again to the shoulder, providing a reservoir for any shrinkage.

5 After about 30 minutes, the resin will thicken, harden and get warm. (Removing the casting while still warm prolongs the life of the mould by reducing burning.) Peel back the mould (this is helped by smearing the outside with washing-up liquid).

6 Surplus resin, poured into the shoulder, can be easily broken off at this stage, saving the time and effort of filing when hard. If tacky, the model or the mould can be cleaned by dipping in a cleaner (acetone). Wash the mould in soap and water, and then dry. The mould must be dusted with powder before re-use.

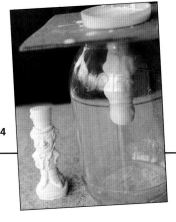

Fig. 3.164

⚠ Safety

Ensure good ventilation in the work area.

Protect against skin irritation with barrier cream and disposable plastic gloves.

Store and dispose of combustible and flammable materials very carefully.

The photos on these pages show the stages in the production of a regular ride. In the first photo you can see the coloured gel coat being applied to a mould of a van. The mould itself is made from GRP that was used to cover the pattern of the van. The pattern can be made from a number of things to give it shape and form (usually chipboard and car-body filler). Unfortunately, the pattern sometimes has to be destroyed in order to remove it from inside the mould. When the colour gel has started to 'cure' (harden) the fibre glass matting and clear resin are applied (clear resin is always used, so that air bubbles can be seen easily and pressed out). On the right, you can see the matting being applied to a mould of a tractor.

GRP – GLASS REINFORCED PLASTIC

This is another forming process which developed from the casting of polyester resin through trying to improve its strength by **reinforcement**. This is achieved by incorporating glass fibre (stranded glass) in polyester resin. Carbon fibre and Kevlar can also be used in woven combinations with glass fibre. Bonded together, the glass fibre forms a flexible mat with a high strength to weight ratio. GRP's tensile strength, impact resistance, and high corrosion resistance, plus the ability to produce complex curved shapes has made the technique suitable for use in a vast range of projects. Car bodies, boat and canoe hulls and caravan shells are examples of industrial products made using GRP. On a smaller scale, models, containers and seating also utilise the process. It combines casting with laminating, taking advantage of the characteristics of the polyester resin (i.e. it can be set without heat or pressure). The contact moulding process demands good hand skills and is suitable for batch production in relatively short runs.

Moulds

The high-quality finish that GRP work can achieve is dependent upon the quality of the mould – the material, in direct contact with the mould, will mirror the slightest defect. It is, therefore, important to consider which face of the moulding needs to be smooth – the **'male'** or the **'female'** (Figure 3.167).

Almost any material and combination, including wood, MDF, plywood, sheet metal, plaster of Paris and GRP, is suitable for mould making. To enable work to be removed from the mould, sides must be tapered or sloped and be free from undercuts. Sharp corners must be avoided, while convex curves provide better structural strength than large flat areas. GRP moulds, made from a master pattern, can be used directly to produce lots of mouldings with a minimum of maintenance. Large and complex split moulds need to be constructed with flanges and are often bolted together. Plaster and other porous materials must be **sealed** before being waxed with **super release wax** and then **polished**. A **chemical release agent** should be applied to the mould and left to dry.

Fig. 3.166 *A pattern used for a tray mould (top), the GRP tray mould, and the final GRP product (bottom)*

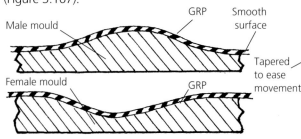

Fig. 3.167 *GRP moulds*

When the sections have cured they are removed from their moulds and trimmed using a diamond-impregnated disc saw and an abrasive wheel. The photo on the left shows a base being removed from its mould, and in the middle you can see a van ride being trimmed. The ride is then drilled, as shown on the right, before the final assembly and incorporation of the drive motor and electronics takes place.

The **lay-up** needs to be carried out in a continuous and methodical manner. Thorough planning and preparation are essential.

⚠ Safety

Avoid skin contact with resin. Always wear appropriate protective clothing.

Work in a well-ventilated area.

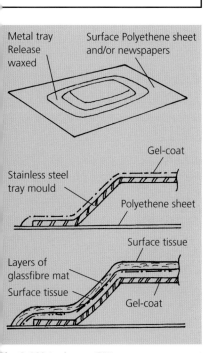

Fig. 3.168 Laying up GRP

Figure 3.168 shows the production of a tray using an existing stainless steel tray as a mould:

1 The immediate work area is prepared and the mould inverted. **Glass fibre mat** is cut slightly over size (matting is available in different weights, i.e. 1oz, 2oz, and finer surface tissue).

2 A **pre-activated gel-coat** (GC150PA), which can be coloured with pigment, is prepared. The **liquid catalyst** 2–4% is added and mixed in. The gel-coat needs to be applied evenly over the moulds' surface with a brush. Aim for a thickness of 0.4mm (or up to 0.6mm, if it is to be used as a mould). This type of resin is **thixotropic** – thick so that it does not run down near vertical faces. NB: The brush must be kept clean, so keep a small container of cleaner (acetone) handy!

3 Give it time to gel (around 15 minutes). The **laying-up** can start when the gel-coat has hardened sufficiently. Check by touching with a finger – your finger should come away cleanly! The **pre-activated lay-up resin** (AP101PA) is catalysed with 2% and brushed liberally and evenly over the gel-coat. The first layer of **surface tissue** is pressed in place and consolidated by **'stippling'** with a brush. It should be thoroughly **'wetted'** through and all **air pockets removed**.

4 This is repeated with subsequent layers of resin and **matting** until the required thickness has been built up. Each layer is worked until completely impregnated with resin. In this case a smoother finish can be obtained on the 'rough side' by using a final surface tissue and stippling against an acetate sheet, which can then be peeled off.

With complex forms, woven tape and rovings can be used to reinforce edges, and ribs and metal inserts incorporated where appropriate.

5 An **exothermic reaction** takes place as the resin gels. When still in the soft rubbery state, known as the **'green stage'**, surplus material can be quickly **trimmed,** using a craft knife. When hard, the tray can be removed from the mould.

NB: The product must be fully **cured** before any machining and finishing operations are attempted. A GRP laminate should reach maturity in about two weeks, depending upon the temperature.

Putting it into practice

1. a) List five personal contributions you can make towards helping to make a workshop a safe working environment.
b) List five rules which should be observed to ensure safe working practice when using any specific piece of machinery (e.g. a pillar drill).

2. a) Explain the term 'datum surface', in relation to marking out materials.
b) Illustrate and state a use for each of the following tools:
    i) a marking gauge
    ii) a mortise gauge
    iii) a scribing gauge
    iv) a pair of dividers
    v) odd-leg calipers.

3. Explain, using a suitable example, what you understand by the term 'jig'.

4. a) Sketch and label the main parts of:
    i) a sash cramp
    ii) any other type of cramp.
b) Why is it neccessary to dry cramp a flat wooden framework prior to gluing up?
c) How would you check for accuracy?
d) What methods of correction would you employ if it fails to be 'square' or accurate?

5. Figure 3.169 shows the cutting ends of three tools used for making cylindrical holes. Account for the differences in shape and, referring to each one, explain why it is particularly suited to its specific use.

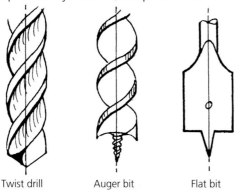

Twist drill   Auger bit   Flat bit

Fig. 3.169

6. a) Illustrate two different methods of mounting metal or acrylic rod, for turning on a lathe.
b) Explain how the lathe would be set up to undertake 'taper turning'.
c) When machining, a cutting fluid is often used. Give reasons for this.

7. A batch of ten aluminium knobs (Figure 3.170) are to be turned on a lathe from a solid round bar which has already been fixed into the lathe chuck. Explain:
a) the first operation to be undertaken
b) how the knurling (diamond pattern) would be made
c) how, using an appropriate turning operation, the 45° chamfer would be produced at the end of the knob
d) how a stopped 5mm diameter hole would be made in the end of the knob
e) the steps needed in tapping a 6mm metric thread in the previous drilled hole.

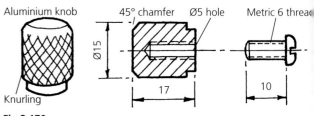

Aluminium knob   45° chamfer   Ø5 hole   Metric 6 thread
Ø15   Knurling   17   10

Fig 3.170

8. A flat-backed house name/number plaque is to be cast in aluminium.
a) develop a suitable design solution
b) construct a flow diagram to explain the sequence of steps, tools and equipment needed to make it.

9. Describe, with reference to tools and equipment, and with the help of sketches, how you would carry out the following processes using acrylic sheet:
a) cutting   **c)** bending
b) drilling   **d)** joining

10. Using a flow diagram, explain the main steps in moulding GRP. Choose a suitable product to illustrate your answer and include all the tools and equipment to be used.

11. Using a flow diagram, show how golf tees are manufactured. Name the processes and suggest a suitable plastic material.

12. Using sketches and brief notes, explain each of the following manufacturing processes, give an example of a product for each of them and a material that can be used with each of them:
a) extrusion   **c)** injection moulding
b) blow moulding   **d)** calendering

13. A lift-off lid is required for the nickel silver container illustrated in Figure 3.171. Wood, metal, plastic or a combination of any of these materials could be used.
a) sketch four possible solutions
b) choose one of your solutions and construct a flow diagram to explain how you would mark out, cut and finish it, giving details of all the materials, tools and equipment you would use.

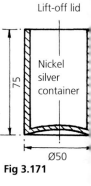

Lift-off lid
Nickel silver container
75
Ø50

Fig 3.171

4·Fabrication

Fabrication is an important forming process by which some products are made (manufactured). It involves fitting together and joining components (parts) sometimes of differing materials to construct an end product, such as a climbing frame or a chair. An example of a large-scale fabricated structure is the windmill (Figure 4.1) which is made up from smaller components that are assembled, fitted and joined together.

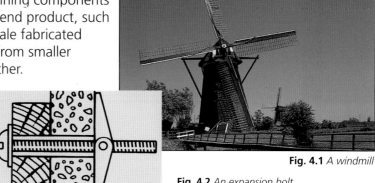

Fig. 4.1 *A windmill*

Many of the ways used to join and hold materials together are to be found in nature. For example, the expansion bolt in Figure 4.2 opens out like the roots of a tree to provide anchorage when it is fitted through a hole in thin flimsy material. This enables the bolt to give support to a heavy object, such as a picture frame hanging from a cavity wall.

Fig. 4.2 *An expansion bolt*

Joints can be made to hold firmly, while still allowing **movement**, just as in the human arm (Figure 4.3) which moves freely. Simple hinging often limits the direction of movement, for instance in the elbow, but the wrist joint, with its complex bone structure, allows for movement in many directions.

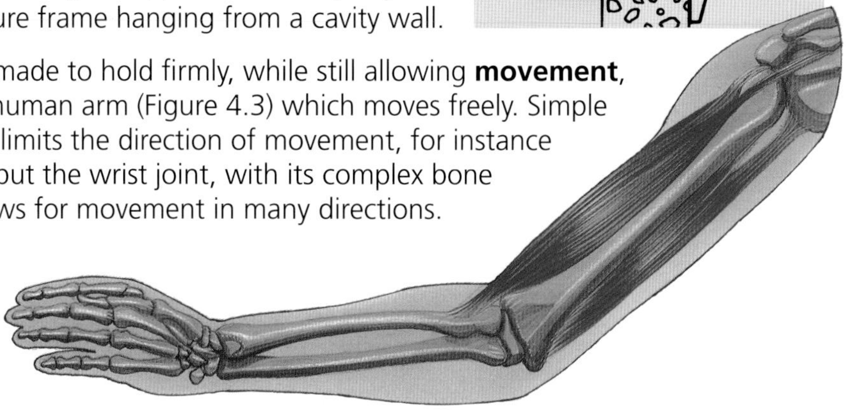

Fig. 4.3
The human arm

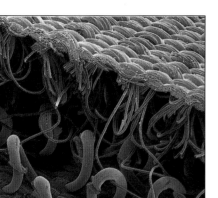

Fig. 4.4 *Velcro's plastic hooks and massed fibres interlock firmly*

When joining materials together the type and nature of the material, and the size and use of the finished product must be taken into account. From early times ropes, made by intertwining creepers, have been used to hold things together. More recently, products such as 'Velcro' have been developed, whose interlocking qualities (shown in Figure 4.4 with its two parts made from plastic hooks and massed fibres) produces a very strong connection when pressed together.

With rigid materials (those which do not bend easily) different ways of cutting and fitting need to be explored. The connecting method used for any particular task depends upon the types of material used. The construction should be designed to exploit and reinforce the strengths of the material and strengthen any weaknesses. Therefore no construction plan can be finalised until materials have been chosen.

To meet functional requirements both **similar** (like) and **dissimilar** (unlike) materials may be used. It must be considered how long joints must last, and whether the connections need to be of a permanent or **temporary** nature. (Enabling things to be taken apart and then reassembled has led to a whole range of **'knock-down'** techniques.)

Joints, whether fixed or moveable, must be able to withstand the forces exerted upon them. Their strength is a crucial factor. **Mechanical connections** and **chemical adhesives**, together with the introduction of extra materials, may be required to achieve successful outcomes. It is important to experiment by trying out ideas. Equally, some form of test procedure is needed to check the strength of jointing and ensure **safety** in all structures, such as the child's climbing frame shown in Figure 4.5.

Aesthetics demands that **appearance** as well as performance is taken into account. Other considerations include economy of effort and resources. By considering these things this chapter attempts to help you develop and use appropriate connections when fabricating.

Fig. 4.5 *Safety must be ensured for all structures*

STRUCTURES

In our everyday world we are surrounded by natural and manufactured structures of infinite variety, ranging from the simple, to the complex.

Manufactured structures make up the built environment, which includes shelter in the form of buildings such as homes, schools and shops; transport in the form of vehicles, roads and bridges; and the provision of energy, through structures such as dams and electricity pylons. Other manufactured structures such as clothing and furniture are designed to improve the everyday quality of our lives.

The natural environment with its plants, animals, birds, fish and insects is a marvellous resource of **natural structures** that have evolved over millions of years. The ways in which structures like flowers, fruit, eggs and shells, contain and safely protect their contents is truly amazing!

Fig. 4.6 *Manufactured goods*

Fig. 4.7 *A natural structure*

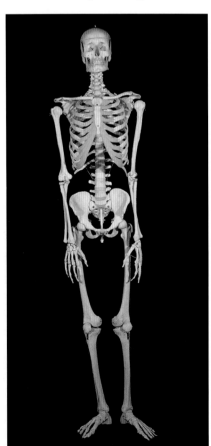

Fig. 4.9 *The human skeleton – a natural frame structure*

Fig. 4.8 *A manufactured frame structure*

Most structures are really **systems**, because they are made up of a number of **parts** (members) held together in their respective positions, that work together as a unit. Often there is a visible form of framework such as in the pylon (Figure 4.8) which is called a **frame structure**, and is also referred to as a **skeletal** structure. The complex skeletal system of the human body (Figure 4.9) is an example. Here the bones, forming the skeleton, provide rigid support and anchorage for muscles as well as protection for the more delicate tissues.

Fig. 4.10 *Drinks cans are shell structures*

Other types of structure, including car bodies and drinks cans (Figure 4.10) have no obvious underlying framework. They are made from the body material itself and are known as **shell structures**. They can be surprisingly strong yet very light, when compared to frame structures.

The soap bubble shown in Figure 4.11, and structures such as eggs, have a skin of material which forms the shape. Here the strength lies in the whole, rather than being shared between the parts. With this type of shell structure, called a **'monocoque' structure**, some areas may be reinforced to give extra stiffness and strength, such as the sub-frame of a car.

All structures must be capable of carrying **loads** and be designed to withstand the **forces** acting upon them without permanently changing shape or collapsing. **Strength** and some degree of **flexibility** are important. The wings of aircraft, for example, need to be able to bend or flex just as trees do. Like all stable structures they must be able to return to their original shape and position when the forces are removed. The type of material chosen for a structure, must reflect this 'elasticity' as well as the overall shape and form.

The school chair shown in Figure 4.12 is a structural system which combines a tubular steel framework with a polypropylene shell in order to support the user. Chairs are designed to support a static load – the weight of the user – which is applied gradually. Chairs are generally not designed for dynamic loading of sudden impact or changes of movement, such as that created by violent rocking or swinging. A dynamic loading results in forces being exerted on the chair that are too great for the structure to support.

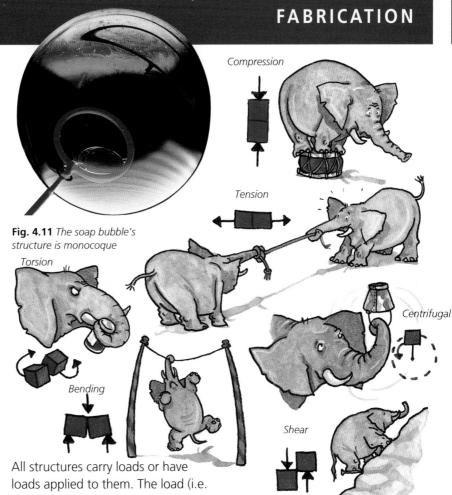

Fig. 4.11 *The soap bubble's structure is monocoque*

Fig. 4.13 *Types of force*

All structures carry loads or have loads applied to them. The load (i.e. the size) of a force is measured in units of force called newtons (abbreviated to N) named after the scientist Sir Isaac Newton who defined force and explained its effect on motion. There are different types of **force** (see Figure 4.13) which can be applied in different directions.

- **Compression** is a push force that tries to squash or shorten. It is applied to a chair when someone sits down.
- **Tension** is a pull force that attempts to stretch or lengthen. Examples are suspension bridge cables and tug-of-war ropes.
- **Bending** is a force that attempts to bend material. It is applied near the middle of a gymnast's bar.
- **Shear** is a sliding force which acts in opposite directions. It is applied to pinions when rock climbing.
- **Torsion** (torque) is a turning, twisting force. It is applied when unscrewing the lid of a jar.
- **Centrifugal** is a spinning outward driving force. It is applied to a spin drier or chemical centrifuge, where it is used to separate solids from liquids.

External forces such as wind, air, gas, water and fluid pressures, apply loads to structures which in turn set up **resisting forces**, creating a complex interaction of atoms. Structures are constructed to distribute and spread loads over more than one place. This distribution of loads together with low centres of gravity helps to ensure stability in such structures as large buildings. An example is shown in Figure 4.14 where the cathedral's flying buttresses enable the walls to remain slender when pierced by windows to allow in more light.

Fig. 4.12 *The school chair combines a tubular steel framework with a polypropylene shell*

Fig. 4.14 *The flying buttresses of Notre Dame*

Beams and bridges

A beam is a bar that carries a load across a gap, such as a fallen tree trunk over a stream which provides a form of bridge. In this case the external forces are inclined to the axis which, being horizontal, means that the load acts vertically downwards as shown in Figure 4.15.

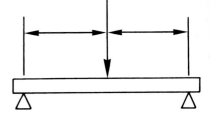

Fig. 4.15 *A beam*

Vertical beams are used in the post and lintel construction found at Stonehenge in Wiltshire (Figure 4.16). Similarly, Greek temples like the Parthenon in Athens (Figure 4.17) use columns as vertical beams. Both illustrate the concept of 'self-weight'. Structures not only have to support loads, but must also support the weight of the material from which they are made. Extra strength is needed to span increasing distances, therefore the stones are heavier and consequently there is an optimum distance between the supporting pillars. The Roman arch (Figure 4.18), built of stone and brick, was a development in the transmission of loads, permitting the span of greater distances.

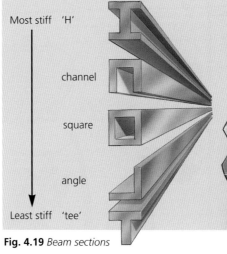

Fig. 4.19 *Beam sections*

In more recent times beam shapes have been made much stronger and also much lighter. Beam sections, as shown in Figure 4.19, are often pierced or castellated to reduce the weight still further. The modern suspension bridge, like the Humber Bridge illustrated in Figure 4.20, is an example of how technology and modern materials can be used to span huge distances. The road deck has been suspended from steel cables which are under tension.

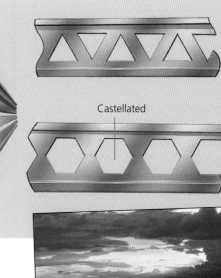

Fig. 4.20 *Humber Bridge (suspension bridge)*

They in turn are supported by concrete towers near the ends of the bridge which are under compression. Design reflects the materials available and is also often closely related to costs.

Fig. 4.16 *Stonehenge*

Fig. 4.17 *The Parthenon*

Fig. 4.18 *A Roman arch*

keystone
forces
large distance

Stability

Stability is a vital consideration for all structures. The structural principles behind stability can be demonstrated using frames built up from flat strips of Meccano, or similar construction kits. In Figure 4.21 **pin-joints** have been made at the corners (nodes) with loose fitting nuts and bolts, which allow movement. You can investigate any possible movement using four, five, or more members joined in this way by holding just one and then applying slight pressure. Try again, using only three strips. This time no movement is possible, except where there is any slackness in the joints.

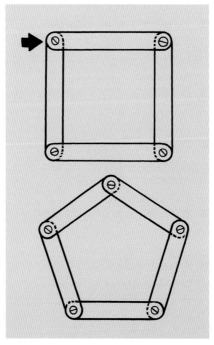

Fig 4.21 *Pin-jointed frames*

The **triangle**, therefore, forms a good basis for practical structures which can be extended by adding one or more additional members (triangulation) as illustrated in Figure 4.22. A frame consisting of identical equilateral triangles, known as a 'Warren girder', has been used to make bridges throughout the world. Similar pin-jointed frames, such as roof trusses, are used in buildings.

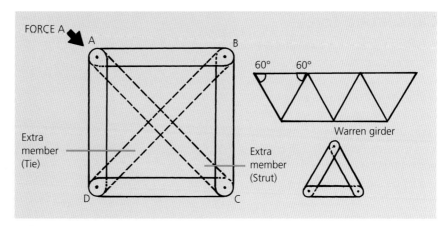

Fig. 4.22 *Triangulation*

The triangular design means that the external roof loads are carried at the joints. Two kinds of force act along the members. Forces acting outwards ↔, trying to stretch and put members under tensile loading, are called **ties**. Forces acting inwards ⟩⟨, trying to shorten and put the members under compressive loading, are called **struts** (see Figure 4.23). Structural members used in tension do not have to be stiff.

In practice, frames can have more bars than are needed for them to be just stiff. Any member in Figure 4.24, such as AC or BD or HJ, which can be removed without affecting the stability is called a **redundant member**. Therefore designers need to consider the most effective and economic use of members when meeting functional requirements.

Fig. 4.24 *Redundant members*

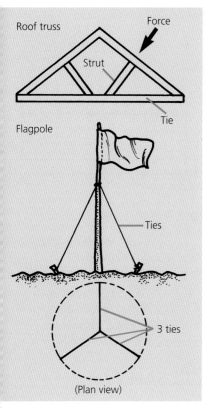

Fig. 4.23 *Struts and ties*

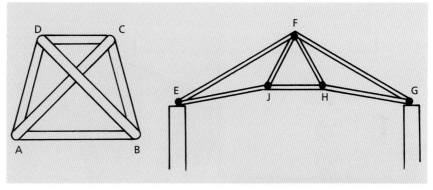

The **cantilever**, shown in Figure 4.25, is a beam supported only at one end, as in the case of a spring diving board over a swimming pool. Brackets for shelves, hanging baskets and some motorway bridges, which meet in the middle, use this type of structure.

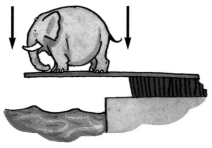

Fig. 4.25 *Cantilever beams*

Fig. 4.26 *The geodesic dome structure of the Vancouver Science World Centre in Canada*

By using geometrical forms, double curved sections can be created to construct structures, such as the **geodesic dome** (Figure 4.26) designed by the American architect Buckminster Fuller.

Calculating forces

In order to make structures that are strong enough but not too expensive, various calculations must be made. Designers and engineers need to calculate the forces that will be acting upon the structure and estimate the load to be carried. Analytical, graphical and mathematical methods of modelling are used, as well as computer simulation. Forces possessing magnitude and direction are called **vectors** and can be represented by a straight line. The overall strength of a structure is dependent upon every component being strong enough to withstand the forces it has to transmit. By identifying each member, Bow's notation can be used to draw up a **force diagram** as shown in Figure 4.27. The construction for finding the **resultant** is known as a **triangle of forces** and is found by measurement. In a **parallelogram of forces** the resultant is achieved by measuring the diagonal.

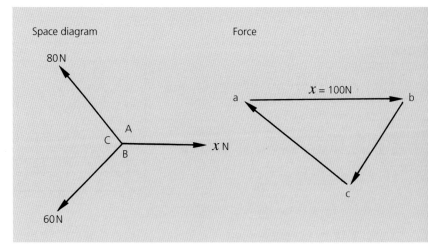

Fig. 4.27 *Space and force diagrams*

In the beam (Figure 4.28) the downward forces acting on it are balanced by upward forces at each end called **reactions**. With the load in the centre, the reaction at X and Y will be equal to half the downward force. If not centrally loaded, the reactions will not be equal.

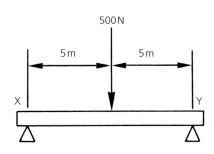

Fig. 4.28 *Beam forces acting*

If the forces are balanced the structure is said to be in **equilibrium**. With the see-saw (Figure 4.29) balance can be achieved by the proper distribution of the load. Using '**moments**', the calculation is shown in Figure 4.30.

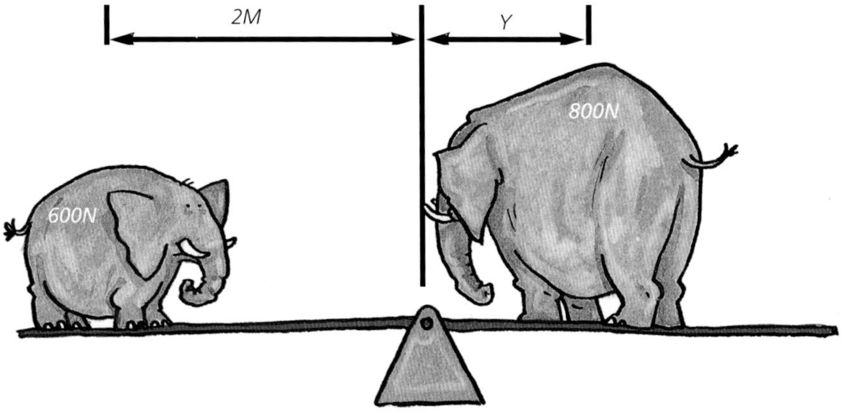

⚠ SAFETY

Calculating and estimating loads is very difficult – overloading lifts, car jacks and children's swings is not uncommon. The scale and magnitude of earthquakes illustrates how loading, as well as force, may cause collapse. **Structural failure** is taken to mean any breakdown, or the onset of permanent strain or fracture. Consequently, structures are designed with a **Factor of Safety**. This is determined by first calculating the forces a structure might have to withstand, and then multiplying this by a factor number, which is often 4, related to the specific Code of Practice. Therefore in the case of a child's swing, calculated to take 120N, the design should in fact be capable of supporting loads of 480N.

Fig. 4.29 *See-saw in equilibrium*

EQUILIBRIUM – CALCULATIONS

Anti-clockwise moments = clockwise moments

See-Saws anti-clockwise moment = 600×2
$$= 1200\text{Nm}$$

See Saws clockwise moment = $800 \times Y = 800\ Y$ Nm
If balanced must be equal
$$\therefore 1200 = 800Y$$
$$Y = \frac{1200}{800}$$

To keep balance, $\qquad Y = 1.5\text{m}$

Second elephant must be 1.5 metres from pivot point.

Fig. 4.30 *Equilibrium calculations*

ADHESIVES

Adhesives fall into the category of **indirect connections** (i.e. connections that involve either some mechanical or chemical device) and are usually of a **permanent** nature. Gluing is rapidly taking the place of other methods of jointing. Technological developments continue, but as yet there is no single adhesive that will join all materials. Many commercial products are available. Which one to use depends largely upon its purpose and the materials to be joined. Some basic understanding of the mechanics of adhesives is helpful when deciding which type is required.

The process of **setting** is important. There are various types of process, which are linked to other factors such as low viscosity, wet surface, low contact angle and low surface tension which also link with the properties of the glues. Setting processes include:

- the drying of a solvent which requires one of the joining surfaces to be porous
- the melting of a solid by heat, then re-solidifying it by cooling (often a reversible process)
- chemical, **secondary bonding**, which is fairly weak
- **primary bonding** where small molecules in liquid form combine to make large molecules in solid form and are influenced by the materials to be glued

Adhesives are either **rigid** or **flexible**. Flexible adhesives are unable to withstand loading and are unsuitable where force is transmitted across the interface of two materials. They are used for 'binding' two things together, where there will be no attempt to separate, such as joining a plastic (Formica) worktop to an MDF (medium density fibreboard) base. On this page and the next you will find a summary of some of the available rigid and flexible adhesives.

Fig. 4.31 *Adhesives*

Scotch glue

Scotch glue, the traditional glue used in woodworking, is made from a gelatine formed from animal bones and hides. It is available in slab or pearl (bead) form which is soaked in water, then heated indirectly using a special glue kettle.

Advantages

It is economical, with no waste because it can be reheated and is ideal for veneering, as it can be softened by reheating.

Disadvantages

It is neither heat nor water resistant. It must be used hot and applied quickly before it starts to cool and gel. Work must be cramped for approximately 12 hours, or overnight.

PVA (Polyvinyl acetate) glue

PVA glue, such as Evostik Resin W, is a popular, widely used wood glue. It is a white ready-mixed liquid, sold in various sized plastic containers. It is also useful for some other materials such as card.

Advantages

Strong and water resistant, it does not stain and has a long shelf life if kept airtight. It requires light cramping and sets in 2–3 hours. Excess glue is easily removed with a damp cloth.

Disadvantages

It is not waterproof.

Synthetic resin glue (Cascamite)

Used on wood, synthetic resin glues set by chemical action and are stronger than PVA adhesives. There are two types: Cascamite and Aerolite 306.
Cascamite is a general purpose wood adhesive of resin and hardener ready-mixed in powder form. It is prepared by mixing with water to the consistency of a thin paste.

Advantages

It is water and heat resistant, and is economical, durable and non-staining.

Disadvantages

It needs to be cramped for 4–6 hours until set. Once set, it is hard on tools.

Synthetic resin glue (Aerolite)

Like Cascamite, Aerolite 306 is used on wood, is a synthetic resin glue set by chemical action and is stronger than PVA adhesives. Aerolite 306 is used for laminating and outdoor work. It consists of two parts: a resin powder and a separate hardener (30% formic acid). It is prepared by mixing the resin powder with water into a thin paste which is applied to one half of the joint. The separate hardener is spread on the other part. Apply the resin with a stick, and the hardener with a bandage brush, not a metal ferrule. Setting only takes place when both surfaces are brought together.

Advantages

It is very strong, waterproof and colourless and has a long shelf life.

Disadvantages

It is not gap-filling and can stain some hardwoods. It needs to be cramped 4–6 hours until set. It sets very hard and blunts tools.

Contact (impact) adhesive

Contact adhesives such as Evostik Impact, Dunlop Thixofix and Time Bond are used for fixing different materials such as plastic laminates, metallic strips and tiles to wood and other materials. Synthetic-rubber based, they require each mating surface to be coated with a thin layer and then left for 10–15 minutes until touch dry. Correct positioning and alignment is essential before bringing surfaces together.

Advantages

They are clean, economical and quick to use with unlike materials.

Disadvantages

They allow little or no time for re-positioning. Good ventilation is essential.

Epoxy resin

Epoxy resin such as Araldite is used to make a rigid bond with unlike materials such as glass, ceramic, wood, metal and plastic (but NOT silicon rubber, polythene or thermoplastic). It is supplied in two parts, as a resin and a hardener. The resin and hardener are mixed together in equal amounts and spread over the surfaces and left to set for 24 hours – though a more rapid setting form is also available.

Advantages

It has good water resistance, insulation and gap filling properties.

Disadvantages

It needs to be spread over a large area to be permanent so high cost prevents its use on large-scale work.

Fig. 4.32 *Using sash cramps for gluing-up work*

Special cements and adhesives

- Suitable cements containing solvents **Tensol 12** and **Tensol 70** are used for joining thermoplastics. The process is shown and explained on p. 129. However, polythene, polypropylene and PTFE cannot be joined in this way.
- Special adhesives containing **Tensol 53** are used for joining PVC. Acetone is used to remove any excess glue.
- **Rigid polystyrene cements** are specially suitable for modelling kit forms such as Airfix. They have the advantage of drying quickly, leaving clear joints. Surplus glue is removed with acetone or carbon tetrachloride.
- **Special DIY all-purpose adhesives** such as UHU are used for many materials except polystyrene foam and most plastics. They have the advantages of being strong and clean, but are not waterproof and are highly inflammable. They can be cleaned up with methylated spirits.
- **Special industrial adhesives** such as Loctite are used as general purpose glues. 'Superglues' bond most surfaces in seconds. They are fast acting but expensive, so are unsuitable for large-scale work.
- **Double-sided tape** adhesive is used increasingly with large flat surfaces such as metal to plastic. It is quick and clean to use but is only semi-permanent.
- **Latex adhesive** such as Copydex is suitable for all types of fabrics, upholstery, paper and card. It is non-toxic.
- **Hot-melt glue** in the form of glue gun sticks is a clear glue used as a general purpose adhesive for sticking different materials. Plastic glue sticks are fed into the heated chamber of the glue gun (Figure 4.33). The melted glue comes out of the gun's nozzle rather like treacle and solidifies very quickly. It sticks almost anything, but is not very strong. It can be messy and chills quickly, giving limited work time.

Hints for using adhesives

- Surfaces to be joined must be clean, dry, free from grease and sufficiently large to give strength.
- Do not apply any type of finish (remove any paint etc. from joints).
- Whenever possible roughen the surfaces to give a **'key'**. This helps the glue to wet the surface more thoroughly.
- Plastics can be de-greased by washing in water containing a liquid detergent, but then allow to dry.
- Select and use the correct adhesive. Read and follow the manufacturers instructions carefully. Ensure enough setting time, during which pressure may need to be applied and a longer curing time allowed, before further work or rough handling is undertaken.
- Assemble joints by dry cramping, and preparing thoroughly the sequential procedure before applying any adhesive.
- Check for squareness and alignment before leaving to set, and where appropriate clean off any surplus adhesive.

> ### ⚠ Safety
>
> Hot glue will burn if touched. Careful supervision is required!
>
> Always note manufacturers warnings! Modern adhesives can be highly inflammable. Ensure plenty of ventilation. Some are solvent based and give off fumes which can be addictive and harmful if inhaled! Keep away from the skin, as contact can cause irritation.

Fig. 4.33 *A glue gun*

MECHANICAL JOINTING

Fabrication (i.e. producing a structure from a number of components) is an effective and advantageous process but particular difficulties arise in joining the parts. This is an area under continuous development, with new, improved and cheaper solutions being produced all the time. The illustrations (Figures 4.34 to 4.37) are only a representative selection of the range of numerous mechanical fasteners available to designers.

Most fasteners require simple provision – often only of the production of holes or grooves. However, the main advantage of mechanical fasteners is their independence of the materials being joined. No mixing of materials or adhesion between the surfaces is required, and standard components can be incorporated into designs with little or no change to the mechanical or other properties as a result of the joining process.

Mechanical connections may be **permanent**, **semi-permanent** or **temporary** (knock-down) and each of these may be **fixed** or **moving**. Hinges (Figure 4.34) generally allow for movement in only one plane, but other devices permit rotary (swing) or sliding movement. Look at a modern kitchen and see how many **special fixings** you can find. The need to access cookers and other appliances for maintenance and repair means that some fixings must be easily removable.

Threaded fasteners (Figure 4.36) are devices where an effective clamping force is produced by an applied torque from screws. They can be clipped, riveted or welded into place. Where material is too thin to have enough threads, plastic expansion insert nuts (clinch nuts) are incorporated, for instance into items like food mixers and refrigerator cabinets.

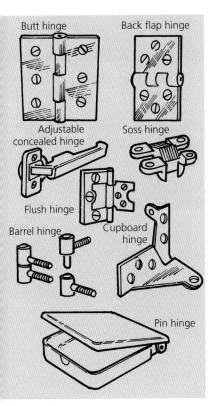

Fig. 4.34 *Hinges made from steel, brass and nylon*

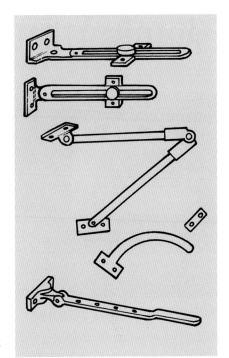

Fig. 4.35 *Stays*

Fig. 4.37 *Clips*

A multitude of catches and clips are made for special applications such as cable retention or for hoses. Figure 4.38 shows a variety of clip fixings, many allowing for disassembly.

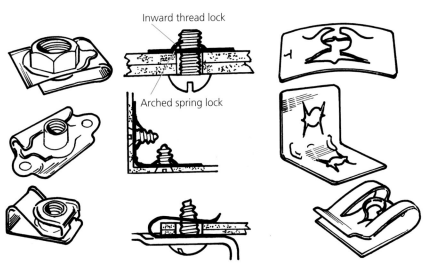

Fig. 4.36 *Threaded fasteners* The diagrams above show the fixings in use

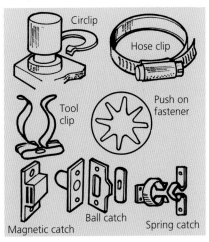

Fig. 4.38 *Special fixings*

From their factory in Immingham (near Grimsby) Auto-Trail Ltd manufacture motorhomes that travel the world. Like so many manufactured products they have to be kept up-to-date with new models coming out. The design and prototyping for new motorhomes is done 'in house'.

In the photo on the left you can see Dave Thomas, who is the Design Director at Auto-Trail. Dave has a degree in industrial design and has specialised throughout his career in transport, working firstly in the car industry and later for coach builders. Dave's sketches and artwork for a new Auto-trail motorhome can be seen here.

Nails, pins and staples

Nails, pins and staples provide a quick and permanent mechanical method of joining wood. Nails grip by friction: the fibres are compressed and forced away from the head of the nail, acting against withdrawal (Figure 4.39). The length of nail is important, it should be approximately three times the thickness of the piece to be fixed.

Nails are generally sold by type, length and weight (quantity). Many different kinds are available made from a variety of materials including steel (galvanized), brass and copper. A representative selection of the most common types is shown in Figure 4.40. (a) **French (wire) nails** are used for general carpentry, frames and pallets. (b) **Lost head wire nails** are used for joinery as the head makes a small hole and is easily punched below the surface. (c) **Oval wire nails** are less likely to split the grain, because the long major axis follows the grain direction. (d) **Panel pins**, used in cabinet work, are fine gauge, small headed and easily punched. They are used for thin materials such as hardboard where little or no punching is required. (e) **Staples** provide a quick method of securing materials.

The square type, used for upholstery, is usually fired in with a staple gun. The heavier round type is used for holding wire and cable. (f) **Corrugated fasteners** are suitable for simple cheap work (i.e. crude butt joints).

Hints for using nails

- Nails can be unsightly, unless hidden by punching (with a nail punch) below the surface.
- When nailing frames, **stagger** the position of the nails as shown in Figure 4.41, to avoid splitting the grain.
- **Pre-drilling** can also prevent splitting.
- **'Dovetail'** nailing in pairs as shown in Figure 4.42 gives extra strength.

Fig. 4.39 *Nails compress the fibres*

Serrations give extra grip
Length
A
B

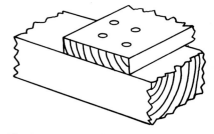

Fig. 4.41 *Stagger nail positions*

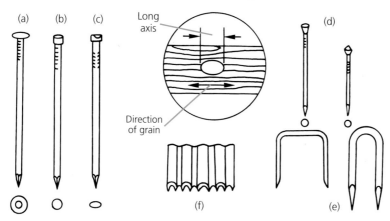

Fig. 4.40 *A range of nails and staples*

(a) (b) (c) (d)
Long axis
Direction of grain
(f)
(e)

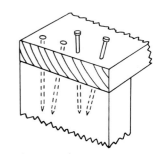

Fig. 4.42 *'Dovetail' nailing*

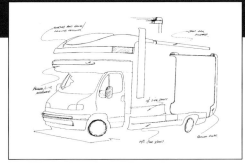

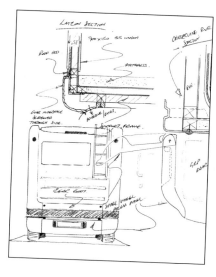

His detail design sketches are turned into prototype motorhomes in the factory.

The detail sketch on the far right is sufficient to act as a working drawing for the manufacture of a prototype. It shows the rear end detail panel for the spare wheel housing and the location of a rear boot. At the top of the sketch you can see the detail of how the roof will be fitted (interfaced) with the cab of a standard Fiat or Peugeot van.

Screws

Wood screws provide a neat and strong method of fixing which can be either **permanent** or **temporary**. They are used to join similar materials, but also metal and plastic to wood. The threads become enmeshed in the grain fibres and, unlike nails, can be easily, removed. Screws made from steel are cheap and strong. Brass screws have a more pleasing appearance and a resistance to corrosion. Finishes include plating with zinc (galvanized), chrome, nickel or black japanning.

You need to specify the **length**, **gauge**, **material** and type of **head** and **slot** when purchasing screws (see Figure 4.43). The screw size (gauge number), commonly 4–10,

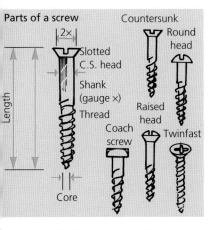

Fig. 4.43 Screws

Gauge No	4	6	8	10
Approx. Diameter	Ø3	Ø3.5	Ø4.5	Ø8

remains the same whatever the length – the higher the number, the thicker the shank diameter. **Countersunk** screws are used where a flush surface is required, and for fixing hinges. **Round head** screws are used when attaching thin metal or plastic fittings such as brackets. **Raised head** screws are for decorative use with door furniture and household fittings, and are often plated. **Twinfast** screws are adapted for such materials as chipboard, having extra and longer threads. **Coach screws** are for securing heavy duty equipment and the head is turned with a spanner.

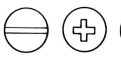

Fig. 4.44 Screwdriver slots

The various types of head require different screwdriver slots – **slotted**, **Phillips** and **Pozidriv** – as shown in Figure 4.44. The Phillips crossed recessed slot, reduces the chances of slipping and the Pozidriv version has an improved grip system.

Screw cups and **caps** can hide or enhance screws, and are useful where frequent removal is necessary.

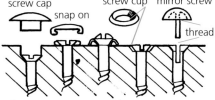

Fig. 4.45 Screw cups and caps

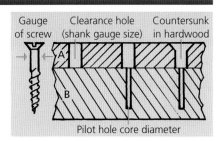

Fig. 4.46 Preparation for fixing screws

Fixing screws

1) Refer to Figure 4.46. Drill a clearance hole, the shank gauge size, in part A.
2) Make a pilot hole, the core diameter, in part B. Drill the hole if in hardwood, or bradawl in softwood.
3) Countersink if appropriate in hardwood.
4) If using brass screws, first insert a steel one, to avoid damaging or twisting-off the head.
5) Lubrication with petroleum jelly, helps prevent corrosion.

Note: Screws fail to grip well in end-grain and some manufactured boards. As shown in Figure 4.47, use either (a) plastic wall plugs, or (b) glue in dowel, to give side grain for better grip.

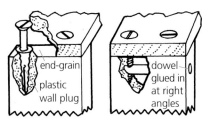

Fig. 4.47 Cut-away sections showing how to secure better grip in end grain

DESIGN PROTOTYPING

Some of the motorhomes made at Auto-trail are not 'custom built' but are based upon van conversions, with the roofs lifted and windows cut into the sides. In these photographs you can see a prototype van conversion taking place. The roof is being fabricated to fit the vehicle. The windows, on the other hand, are standard parts that are 'bought out' from other manufacturers. When the van has been converted to its full potential as a motorhome it can be evaluated.

The photograph on the far right shows a 'Styrofoam' and wood model of the rear panel of the new design shown on the previous page. From this model a GRP mould will be taken and the prototype motorhome will then have a GRP rear panel made from this mould.

Machine screws, nuts and bolts

Machine screws, nuts and bolts provide a convenient method of securing parts. Easily undone, this **temporary** method of fixing is suitable for all kinds and combinations of material. **Machine screws** (set screws) are available in a wide range of diameters, lengths, materials, head shapes and thread forms. The modern **ISO metric thread**, available in a coarse or fine series, is an attempt to standardise past thread systems such as **BSW**, **BSF**, **UNC** and **BA**, many of which are still in use. Various types of head are available (see Figure 4.48). Grub and socket-head screws are used with moving (revolving) parts for safety reasons.

Bolts made from high tensile steel are mechanically strong. Square or hexagonal headed, they are threaded for all, or just part of their length.

Fig. 4.49 *Types of nut*

Nuts must be **matching** in diameter size and thread form. They range from wing nuts made for easy removal and standard hexagonal nuts to special locking nuts which resist coming loose (see Figure 4.49).

Washers (Figure 4.50) protect the surface when nuts are tightened, spreading the load and preventing loosening caused by vibration.

Self-tapping screws (see Figure 4.51) are suitable not only for thin sheet metal, but also plastics. Made from hardened steel, they cut their own thread as they are screwed in. Preparation requires a clearance hole and a pilot hole, equal to the screw's core diameter.

Fig. 4.50 *Washers*

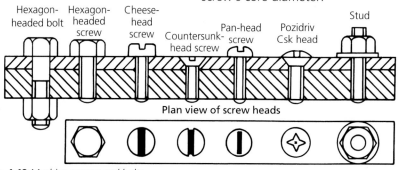

Fig. 4.51 *Self-tapping screws*

Fig. 4.48 *Machine screws and bolts*

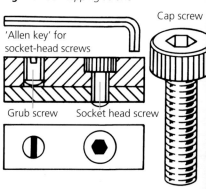

MOTORHOME PRODUCTION

Manufacturing motorhomes is a labour-intensive business, with many hours of work needed on each vehicle. The starting point is a van chassis which has to be extended by welding on extra chassis members in order to accommodate the extra space required. Compare the photograph of the chassis on this page with the photograph of the finished motorhome and see how much further it needs to extend past the rear wheels. Not all vans are built on a chassis construction – some are monocoque structures with no chassis.

Motorhomes are essentially a wooden frame construction with a plastic laminated plywood skin on the inside and sheet aluminium on the outside. The wooden frame is laid out very accurately on a large table and staple guns are used to fix the frame corners. The frame is then covered with PVA glue and the sheets of ply are laid on. These too are stapled (see left). Once the glue has bonded it forms a rigid structure and the plywood can then be trimmed to shape around the frame.

Making screw threads

Screw cutting is appropriate for metal and plastics. ISO metric threads (M4–M10) will satisfy most project needs.

Tapping

The term 'tapping' is used to describe cutting an internal (female) thread. A set of three taps (see Figure 4.52) are used in sequence, with a **tap wrench**. The taper tap makes for easier starting and in thin material may itself give a full thread. The tapping hole must be drilled to the correct size – refer to the chart of tapping sizes in Figure 4.53.

The size of the hole should be smaller than the nominal size of the screw to allow for cutting the thread. For example, an M6 thread needs a 5.0mm diameter hole.

Nominal Diameter	Pitch (mm)	Tapping drill size (mm)
M4	0.7	3.3
M5	0.8	4.2
M6	1.0	5.0
M8	1.25	6.8
M10	1.5	8.5

Fig. 4.53 *Tapping drill sizes*

The cutting sequence involves turning clockwise half a turn and then anti-clockwise a quarter turn to break the swarf. Remember, high speed steel (HSS) taps are brittle and extra care must be taken with blind (stopped) holes, shown in Figure 4.54. The tapping hole needs to be deeper than the required depth of thread to accommodate any debris and the chamfered end of the plug tap.

Threading

Threading describes the process of cutting the external (male) thread. A **split die** is held in a **die stock**, shown in Figure 4.55. Three screws locate and provide adjustment for the die. It fits into the stock with the size information clearly visible. This enables the tapered lead of the die to ease starting. Chamfering the end of the nominal rod (Figure 4.55b) also helps. Tightening the centre screw opens the die to give a tight thread. Tightening the two outer screws will close the die, making a fuller thread, but looser fit. Dies are cut in the same sequence as taps.

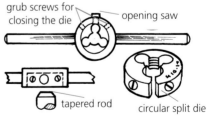

Fig. 4.55 *Split die and die stock*

Hints for screw cutting

- Check and test the alignment when starting taps and dies.
- Use a lubricant such as oil or cutting compound, or use paraffin for aluminium. (Brass and plastics are self-lubricating.)
- Accurate screw cutting can be undertaken, or started, using a lathe.

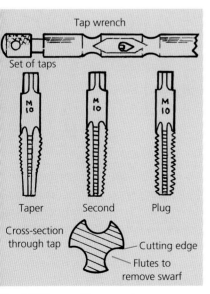

Fig. 4.52 *Taps and tap wrench*

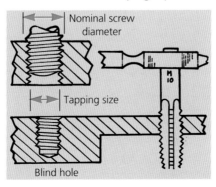

Fig. 4.54 *Tapping*

When the plywood has been trimmed, the wooden frame, with its interior skin forming the motorhome side panel, is then passed through a spray glue process. The side panel is laid on a roller table and is passed through a machine which applies the adhesive. The specially manufactured adhesive which will bond the panel to the aluminium outer skin is applied through a moving spray head. The process is enclosed so that fumes can be extracted.

The panel is then transferred to a vacuum press table where the aluminium sheets that form the outer skin are arranged and located on it. The aluminium is pre-painted with a paint that has been baked on the surface. The panel is covered for protection and a heavy gauge plastic sheet is clamped over it. The air from under the sheet is then evacuated and the aluminium is sucked tight into the panel whilst the glue is curing. The three photographs on the left show the stages in this process.

Knock-down fittings

This type of fastening offers opportunities to make self-assembly products, especially large items such as kitchen, bedroom and other furniture. For such products, supplied in flat-packs, **knock-down (k-d) construction** provides the means to dismantle and re-assemble whenever necessary. All screw assemblies and mechanical k-d fittings are especially suited for use with modern manufactured materials such as MDF or Conti-board.

Line (frame) connections are of five basic types, as illustrated in Figure 4.56.

Corner blocks (see Figure 4.57) comprising of one- or two-piece plastic block-joints, enable light unit construction such as shelving to be made from a limited number of components.

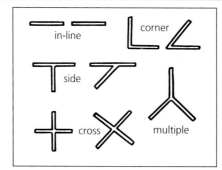

Fig. 4.56 *Types of connection*

Mixing and combining multi-materials in items such as tables and chairs creates interesting problems and varied solutions. For instance, the functional part, the construction, is often seen rather than hidden. Also, tubular steel and aluminium legs, used in combination with wood rails require some form of **frame connectors** – the aluminium barrel and Allen cap machine screw provides the solution (see Figure 4.58). Location pegs, or dowels, are employed to prevent the rail rotating.

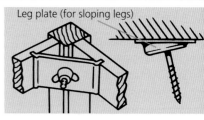

Fig. 4.59 *Leg fastenings*

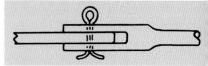

Fig. 4.60 *Cotter (split) pin*

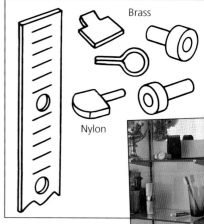

Fig. 4.61 *Shelving fitments*

Figure 4.59 shows leg fastenings, used where joints need to be frequently taken apart. Simple **cotter (split) pins** (Figure 4.60) make effective fastenings where freedom of movement in the joint is required. Shelving fitments (Figure 4.61) also offer a wide range of solutions and techniques.

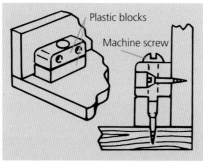

Fig. 4.57 *Corner block joints*

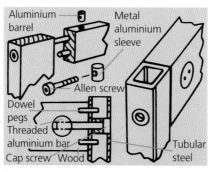

Fig. 4.58 *Frame connectors*

Trimming the excess aluminium sheet from around the edges of the side panel is carried out using a pneumatically powered (compressed air) nibbling tool, which you can see in the photo on the far right. This has a shearing action that cuts through the thin aluminium very quickly. Pneumatically powered tools are commonly used for many hand assembly and fabrication processes.

Compressed air provides plenty of power for hand tools, it is safer than electricity and the tools tend to be more robust and have a longer life.

Other pneumatic hand tools seen in use at Auto-Trail include screwdrivers, staple guns and disk sanders. Electrically powered hand held routers are used for cutting apertures for windows etc., in both laminated plywood and aluminium.

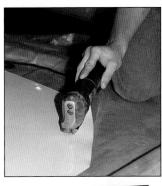

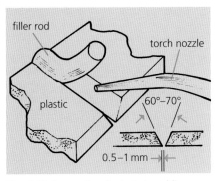
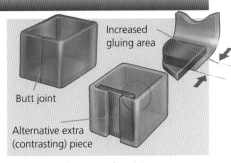

JOINING SIMILAR (LIKE) MATERIALS

Joining plastics

Plastics can be drilled, tapped and joined by means of screws and fastenings. **Welding** is also possible with suitable non-inflammable **thermoplastics**, and comparisons can be made with welding metals. Two or more pieces of the same plastic material are joined using heat to give a joint of similar composition. There are various techniques. **Heated-tool welding** is used to seal plastic film or sheeting such as polythene bags. The material is softened by a heated tool (even a soldering iron can be used) and then pressed together until it cools and solidifies. A special tool, with a heated roller, will give uniform strength and a better finish.

In **hot-air welding** hot air is applied with a torch, heating the surfaces to be welded and the filler rod. As shown in Figure 4.62, a space is left between the surfaces and the softened filler rod lies in the junction. Industrial methods of welding include **high frequency welding** to ensure quality joints for containers, clothing seams and soles of shoes.

Fig. 4.62 *Preparation for hot-air welding*

Hints for welding plastics
- Practise on scrap material to develop technique and expertise.
- Temperature is important and varies with the plastics' properties.
- Dissimilar materials, such as PVC and polythene cannot be welded.

The most popular method of joining **acrylic** is to use a special adhesive (Tensol cement). **Tensol 12** is solvent based and interacts with the surface of the acrylic before evaporating. Supplied ready for use, it sets quickly, but is not very strong. **Tensol 70** is a two-part adhesive that cures by chemical action, leaving joints almost as strong as the acrylic itself. Careful mixing is required in accordance with the manufacturer's instructions. It will remain usable at room temperature for up to 20 minutes, with a setting time of about 1.5 hours.

Fig. 4.63 *Surface areas for gluing*

The size of gluing area directly effects strength. Try to overlap and join away from bent corners (see Figure 4.63).

Hints for gluing plastics
- All joints must be thoroughly prepared (see Figure 4.64).
- Joining surfaces must be clean and given a key with wet and dry paper.
- Use masking tape to protect the immediate finished surfaces from being spoilt by excess adhesive.
- Support the structure firmly, using limited pressure until set.

> ⚠ **Safety**
> Ensure good ventilation when you are using glue.

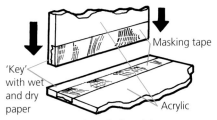

Fig. 4.64 *Preparation of a butt joint*

Many of the component parts for Auto-Trail motorhomes are from outside suppliers, including plumbing fittings, electrical fittings and several that could be found in a car or caravan. It is always an advantage to be able to use standard items where possible, but many things do have to be 'custom' made. Most large fittings are 'bought out' – fridges and window blinds from Germany, toilets from Holland, and ovens and water heaters that are made in the UK. The carcasses for the internal cupboards and wardrobes are made in the factory to suit the vehicle but the doors for these are standard caravan doors.

On the far right you can see the roof insulation being applied. All of the spaces between the inner and outer skins are filled with heat insulating materials. The aluminium sheets for the roof can be seen in the other photograph, they are supplied to the factory in pre-painted roll form and are screwed on to the top of the side panels.

Joining wood

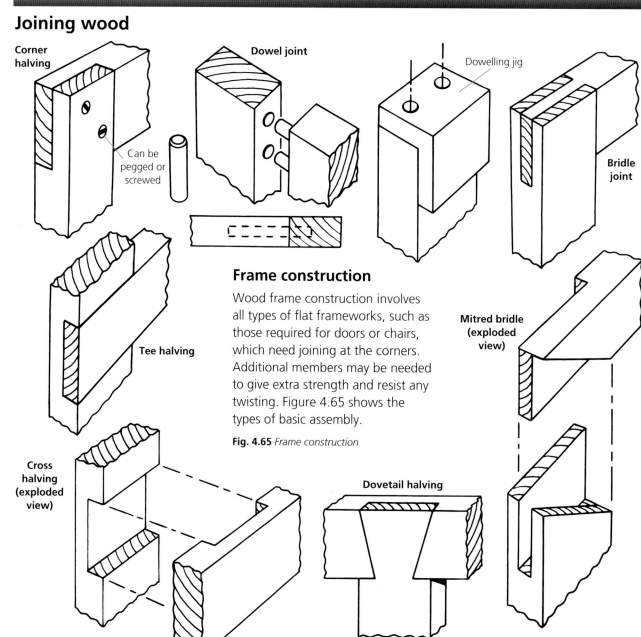

Corner halving

Dowel joint

Can be pegged or screwed

Dowelling jig

Bridle joint

Tee halving

Frame construction

Wood frame construction involves all types of flat frameworks, such as those required for doors or chairs, which need joining at the corners. Additional members may be needed to give extra strength and resist any twisting. Figure 4.65 shows the types of basic assembly.

Fig. 4.65 Frame construction

Mitred bridle (exploded view)

Cross halving (exploded view)

Dovetail halving

The finishing processes include the application of 'trims' over all of the corners and joints – plastic for internal joints, aluminium for external. To the bottom right you can see a transfer being applied. Transfers are added to those panels that are not pre-finished, such as the doors. These are 'sticky back' but the surfaces to which they are applied must be absolutely clean and free from any grease or oxides – even touching the surface with hands will spoil the adhesion. The internal upholstery, carpets and curtains completes the fabrication and after a final inspection and electrical and mechanical check Auto-Trail motorhomes are given a final clean ready for dispatch to the customer.

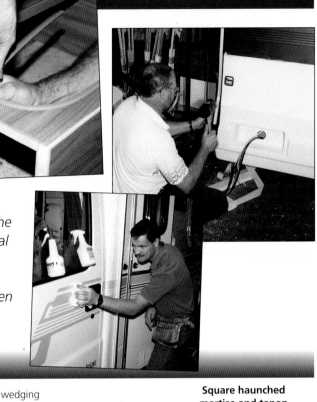

Traditional methods of joining materials involve **interlocking** to give strength. This means that the load is spread directly through the material and over a much greater area of the joint. Adhesives assist by preventing the parts from separating. Figure 4.66 shows variations within one particular family of joints – **mortise and tenon**. All require accurate marking out and precise cutting.

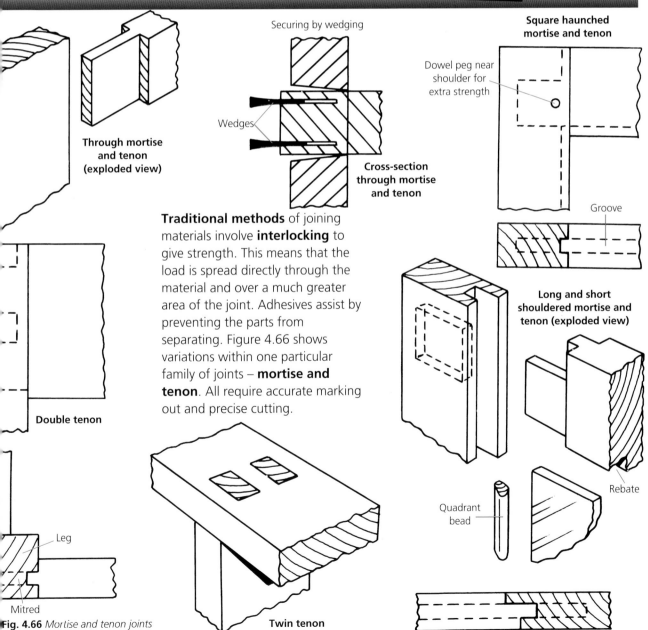

Through mortise and tenon (exploded view)

Securing by wedging

Wedges

Cross-section through mortise and tenon

Square haunched mortise and tenon

Dowel peg near shoulder for extra strength

Groove

Double tenon

Long and short shouldered mortise and tenon (exploded view)

Rebate

Quadrant bead

Leg

Mitred

Fig. 4.66 *Mortise and tenon joints*

Twin tenon

Caravans are an interesting fabrication to study – they are built from the bottom upwards, starting with the fabricating of the caravan chassis. At Vanroyce Caravans in Immingham, the main chassis members are 'bought out' from outside suppliers. In the photographs on this page you can see the pressed steel longitudinal and cross members, and the axle units. The steel is galvanised (given a coating of zinc) so that it resists corrosion. The chassis is bolted together on a jig that ensures correct alignment, holds the members together and also rotates to allow access (see the photo on the next page). Whilst still in place on the jig the plywood floor is fixed to the chassis and is pre-coated with a weather resistant facing. After this, the carpet is laid, which you can see being done in the bottom photo on the next page.

Box construction

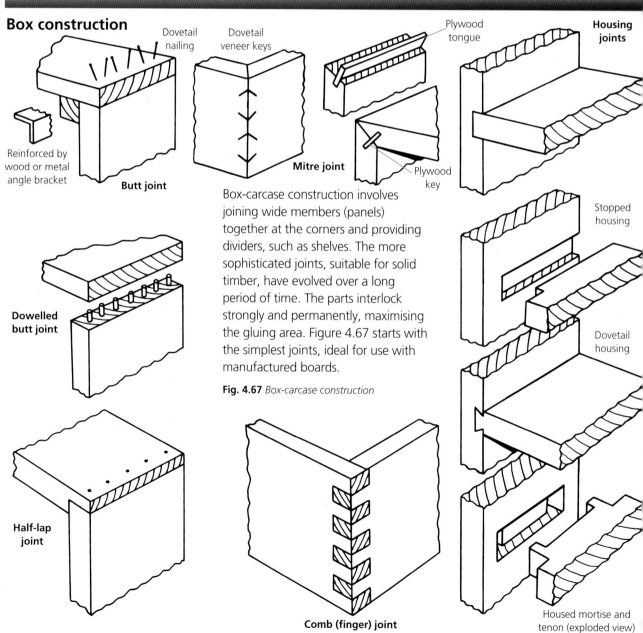

Box-carcase construction involves joining wide members (panels) together at the corners and providing dividers, such as shelves. The more sophisticated joints, suitable for solid timber, have evolved over a long period of time. The parts interlock strongly and permanently, maximising the gluing area. Figure 4.67 starts with the simplest joints, ideal for use with manufactured boards.

Fig. 4.67 *Box-carcase construction*

Dovetail nailing

Dovetail veneer keys

Plywood tongue

Housing joints

Reinforced by wood or metal angle bracket

Butt joint

Mitre joint

Plywood key

Dowelled butt joint

Stopped housing

Dovetail housing

Half-lap joint

Comb (finger) joint

Housed mortise and tenon (exploded view)

This is much easier to do at this early stage. However, it does then need to be covered with a plastic sheet to protect it from being damaged.

The sides are then erected and most of the internal fittings added. The sides of a caravan are built around a wooden frame – the inner skin is made from plastic-laminated plywood. The empty space between the two skins of the wall is filled with insulating materials. The details of this type of construction are shown in the 'Auto-Trail' examples earlier in this chapter.

Blinds being fitted to the windows using a pneumatic screwdriver

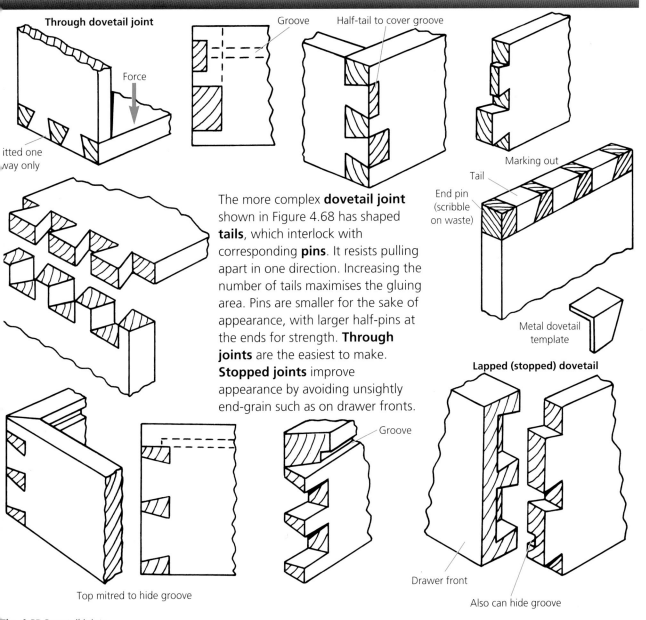

The more complex **dovetail joint** shown in Figure 4.68 has shaped **tails**, which interlock with corresponding **pins**. It resists pulling apart in one direction. Increasing the number of tails maximises the gluing area. Pins are smaller for the sake of appearance, with larger half-pins at the ends for strength. **Through joints** are the easiest to make. **Stopped joints** improve appearance by avoiding unsightly end-grain such as on drawer fronts.

Fig. 4.68 *Dovetail joints*

In the photograph on this page you can see two men working on a caravan roof at Vanroyce Caravans. They are locating the wires for the low voltage lighting circuit. All of the wiring must fit between the inner and outer skins. These particular caravans have two lighting circuits – one that runs on 12v, for using with a battery, and another that runs from the mains when the caravan is 'hooked up' to a supply.

You can see the complex shaping of the wooden battens and the plywood panels on the front and the roof. These are all pre-cut to shape and part-fabricated before the final fabrication takes place. You can see also that most of the interior cupboards and fittings are in place. This is done before the back and the roof go on. There is not a lot of room to work inside a caravan so it is important to plan the fabrication procedure very carefully.

Riveting

Riveting is a quick and convenient method of fixing two or more pieces of material together in a **permanent** fashion. It is ideally suited to industrial application, providing a cheap alternative to threaded fastenings. The choice of rivet depends on its use. Rivets can form hinge pins in moving joints, or rigid joints in sheet material. Traditionally used with sheet metal, this method is increasing applied to plastics and even wood. Rivets are classified by length, diameter, material and head pattern. A wide range of special rivets is available (see Figure 4.69). They are made from soft, easily deformed materials such as iron, aluminium and copper, as well as low-carbon steel and stainless steel.

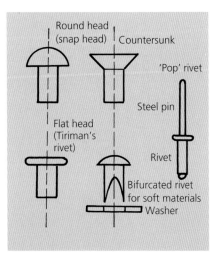

Fig. 4.69 *Types of rivet*

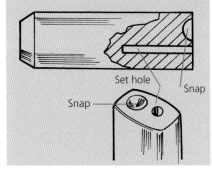

Fig. 4.70 *A combined set and snap*

Conventional riveting employs a **set and snap** (see Figure 4.70). The flow diagram (Figure 4.71) illustrates the process. Countersunk heads are used where a smooth surface is required. A good tip is to drill all the holes in one plate and only one in the other, before securing with a rivet, then proceed by further drilling and subsequent riveting.

Fig. 4.71 *Riveting technique*

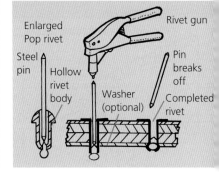

Fig. 4.72 *Pop riveting*

Pop rivets are quicker to use than ordinary rivets and can also be put in from one side, but they do not give such a strong joint. They consist of a hollow rivet mounted on a head pin. In application (see Figure 4.72) the pin is gripped in the rivet gun and the rivet placed in a pre-drilled hole. Squeezing the gun withdraws the pin. At a pre-determined tension it breaks, leaving the formed rivet with the pin head retained in it. Sometimes a washer is used at the back to support the rivet, especially with softer materials.

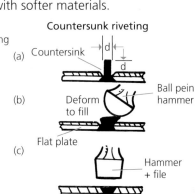

The roof, the front panel and the rear panel are made from GRP. These components are supplied to Vanroyce by a specialist manufacturer. They have to be very accurately made so that the GRP panels fit. In the photographs on the right you can see a front panel, which will be fixed on to the caravan that is shown with the rear GRP panel in place and awaiting its roof.

Notice on the GRP panel how the location areas for the manoeuvring handles are reinforced. These handles are fixed through the GRP into the wooden framework so that the GRP is not stressed and cracked when the caravan is being pulled out.

The finishing touches – the lounge interior of a Vanroyce caravan

Joining metal

Heat processes are used for fabricating metal to give permanent results. Soldering is a means of joining metal surfaces by melting an alloy, with a lower melting point, between them. The alloy unites the surfaces, forming a joint as strong as the alloy itself.

Soft soldering

Soft soldering is a quick method of joining most metals such as copper, brass, tinplate and steel. An exception is aluminium. The process is best confined to light fabrication where joints are not subjected to heat and vibration and so do not need to be very strong. Soft solder is an alloy made from varying proportions of tin and lead with antimony. The melting point varies according to composition, ranging from 183° to 250°C. The solder used for electronics contains more tin than lead, making it flow more easily at a lower temperature. The solder for tinplate or plumbing copper waterpipes contains more lead than tin. It melts at a higher temperature and sets harder.

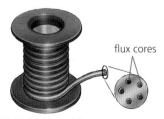

flux cores

Fig. 4.73 *Multi-core solder*

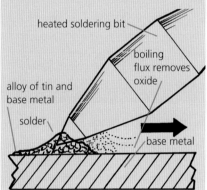

heated soldering bit

boiling flux removes oxide

alloy of tin and base metal

solder

base metal

Fig. 4.74 *The soldering process*

In preparation for soldering the joint surfaces must be clean. Use a suitable abrasive (e.g. emery cloth or steel wool) and avoid touching the area. **Fluxes**, available as liquids, powders or pastes, have been specially developed to protect the cleaned surfaces from oxidisation during heating; solders only stick to clean metal. The flux also helps the molten solder to flow freely by breaking down surface tension. **Active fluxes** (e.g. bakers' flux) contain zinc chloride which chemically clean the surfaces. However, it is highly corrosive and must be washed off immediately upon completion. **Passive fluxes** are non-corrosive, but they only protect and do not actually clean. An example is the multi-core solder (Figure 4.73) used in electronics, which has cores of resin flux running throughout its length.

Apply solder sparingly

Fig. 4.75 *Heat and solder*

In the joining process close-fitting joints are essential to ensure that the capillary action unites the surfaces (Figure 4.74). There are several ways of applying the necessary heat and solder. An electric soldering iron is cleaned, while hot, using a wet sponge and then 'tinned' with a thin film of solder. Heat and solder can be applied together as shown in Figure 4.75, or **sweated**, by tinning both parts of the joint first as shown in Figure 4.76.

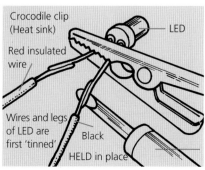

Crocodile clip (Heat sink)

LED

Red insulated wire

Wires and legs of LED are first 'tinned'

Black

HELD in place

Soldering iron applied under the joint area

Fig. 4.76 *Sweated joints*

Within the GKN Group of companies many specialised metal fabricating processes take place. These range from manual welding using electric arc and MIG techniques through to robot-controlled resistance, MIG welding and very specialised techniques such as friction welding.

Vehicles such as the traditional 'Land Rover Defender' have been in continuous production for many years although the numbers of vehicles produced is not high. For this reason, and the fact that the vehicle can have around 30 variations, investment in some of the newer high volume manufacturing technologies has been unnecessary.

In the photograph above you can see hand welding taking place on 'Defender' chassis members. Both conventional arc welding, such as that described on page 139, and MIG welding are used. The MIG (metal inert gas) welding process uses a continuous wire feed as both the electrode and the filler rod.
The transfer of the welding arc and the metal from the wire electrode to the work piece takes place within a shield of inert gas, usually argon. This shield prevents surface oxidisation and the formation of slag.

The new Range Rover chassis is also built by GKN, at their Telford Engineering Products factory. A special rotating welding jig is used to hold the chassis components in position whilst welding takes place. The jig, which you can see in the photo on the right, enables the chassis to be very easily tipped at any angle to facilitate the welding process.

Hard soldering

Hard soldering is much stronger than soft soldering, but requires higher temperatures. Soft solders melt at around 200°C, whereas the lowest melting point of hard solder is 625°C. The principle of local alloying and using a flux remains the same. The extra heat requirement is supplied by using a gas/air torch.

Silver soldering

Silver soldering is so called because hard solder contains silver alloyed with copper and zinc, giving melting points ranging from 625°C to 800°C. It enables work to be joined in several stages, first using solder with a high melting point, working through lower melting points to finally the lowest, called 'easy-flo'. This avoids the risk of earlier joints coming apart when applying heat for the later ones (see Figure 4.77).

Proper **joint preparation** (Figure 4.78) is important and thorough cleaning is necessary with the application of an active flux.

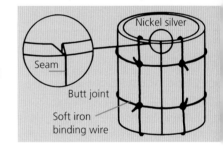

Fig. 4.78 *Preparation for silver soldering*

A special '**easy-flo**' **flux** is used for the lowest melting silver solder, whilst medium and hard grade solders use a borax flux.

Pre-heat the joint with a gentle flame, then concentrate to a small hot flame to achieve a dull red heat. Solder flows to the hottest part, following the flame along the line of the joint.

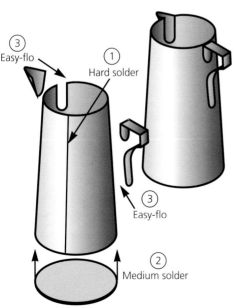

Fig. 4.77 *Fabricating a jug in stages*

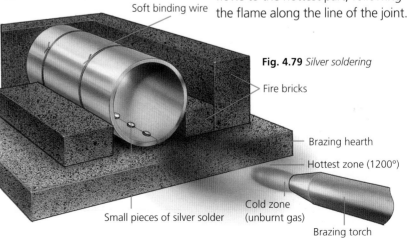

Fig. 4.79 *Silver soldering*

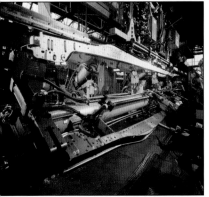

Robot spot welding

Welding can be a repetitive process and is often carried out in an unpleasant environment. These factors make it an ideal process for using robots. Robots are an expensive investment and are therefore more suited to high volume production. In the photograph on this page you can see robots spot welding chassis members for Ford Transit vans and minibuses. Each member has 250 spot welds and 8000 chassis are produced every week. Spot welding is a resistance welding process. Electric current passes between two copper electrodes through the material to be joined. At the point of highest resistance (the joint line) localised heating takes place. Pressure is then applied by the electrodes and a weld nugget is formed.

Robots are also involved in MIG welding. The components to be joined in this way must be very precisely aligned and accurately located – robots are not able to make judgements and allowances for variations like a human operator could.

The spot welding process

copper electrode — pressure

weld nugget — heat-affected zone

pressure

Robot MIG welding

Brazing

Brazing is a technique similar to soldering, except that considerably higher temperatures are needed. Brazing **spelter** is an alloy of copper and zinc (brass) and melts in the range 870°–880°C. This results in a much stronger joint, since brass is stronger than solder.

Again, an air blown (brazing) gas torch (Figure 4.80) is needed to maintain a hot flame. This does create a limiting factor, making it too hot to use with brass and copper, but it is ideal for mild steel.

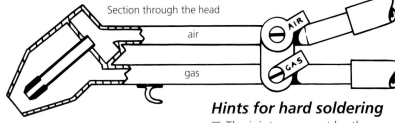

Section through the head

air

gas

AIR

GAS

Fig. 4.80 *Brazing torch*

Joints do benefit from interlocking (Figure 4.81), but they should all be wired or held securely to allow for expansion during heating. Use a flux with borax or a proprietary brand like Sifbronze.

Hints for hard soldering

- The joint area must be thoroughly clean and fluxed.
- Allow time for spelter or solder to flow, melting on the hot metal (not in the flame).
- Pre-heat gently, avoid too fierce a flame which might blow away flux and spelter or solder.
- Surround with fire-bricks to reflect all possible applied heat.
- Heavy sections will require the most heat, at least to dull red.
- Maintain the heat until the spelter or solder flows throughout the joint.

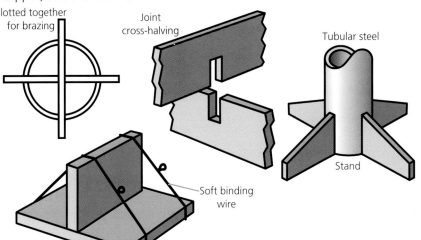

Slotted together for brazing

Joint cross-halving

Tubular steel

Stand

Soft binding wire

Fig. 4.81 *Joint preparation for brazing*

⚠ Safety

Extremely high temperatures are involved, so welding should only be carried out under close supervision.

When a company like GKN is involved in high volume production large investments have to be made in manufacturing. The cost of setting up a fully automated robot welding cell for the production of Ford Escort front cross members was £6 million! You can see a photograph of it on the right. The cell and the robots within the cell are controlled by PLCs (programmable logic controllers). These are a type of computer/processor that is dedicated to a specific task, and is also able to communicate with other PLCs and computers.

Component parts, made from sheet steel that has been accurately formed using transfer press forming (see page 88) are loaded into the cell by a parts-handling robot. The spot welding necessary to fabricate the cross member is carried out by multi-headed spot welding robots that are able to position themselves around the components at very high speeds. The component leaves the cell via a conveyor system that takes it on to the next manufacturing cell.

Welding

Welding offers a permanent method of fastening and fabricating products from a wide range of materials. Welding is the joining of two materials (usually metal) in their liquid form which solidifies and fuses together to form a joint that is as strong as the parent metal. Industrially there are many ways of achieving this fusion.

Within school workshops two basic methods of welding metal are possible: oxy-acetylene and electric arc.

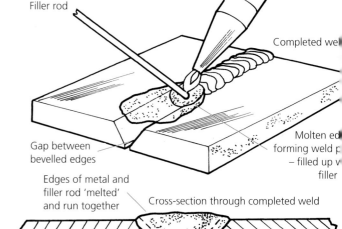

Filler rod

Completed we[ld]

Gap between bevelled edges

Molten ed[ge] forming weld p[ool] – filled up v[ith] filler

Edges of metal and filler rod 'melted' and run together

Cross-section through completed weld

Fig. 4.83 *The oxy-acetylene welding process*

In **oxy-acetylene welding** a heat source of around 3500°C is produced by burning acetylene gas in oxygen. Fine adjustment to the ratio of gases is made on the hand-held blowpipe. Excess oxygen gives the hottest flame, but a **neutral flame**, with equal volumes of gas, is the most widely used. Adjustments are made to suit the thickness and type of metal being welded. During the welding process (Figure 4.83) a pool of molten metal is created. A **filler rod**, of the same metal as that being joined, is dipped into this and melts, filling the joint. Fluxes are used with some materials, but not steels.

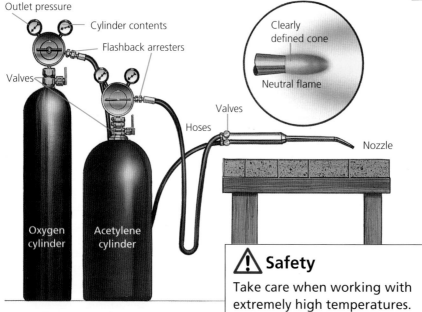

Outlet pressure

Cylinder contents

Flashback arresters

Valves

Valves

Hoses

Nozzle

Clearly defined cone

Neutral flame

Oxygen cylinder

Acetylene cylinder

⚠ **Safety**

Take care when working with extremely high temperatures.

Fig. 4.82 *Oxy-acetylene equipment*

An interesting and very specialised welding process that has developed over recent years is 'friction welding'. This process involves rotating a component at very high speed whilst it is in contact with the component to which it is to be welded. The resulting friction between the two components raises the temperature at the point of contact to a level where the metal is molten. The two parts are then forced together under pressure raising the temperature enough for welding to take place. The process has many advantages over conventional welding techniques – it is very fast, the heated area is localised and the rotation very effectively removes any oxides and surface irregularities.

The photograph on the top right shows part of a drive shaft for a car where the forged 'cup' shape in the end is friction welded to the shaft, after machining there is no evidence of any joint.

On the bottom right you can see a mounting point being friction welded to the side of a 'Warrior' special purpose vehicle. The material is aluminium, often a difficult metal to weld. The machine being used was developed specifically for this very specialised task.

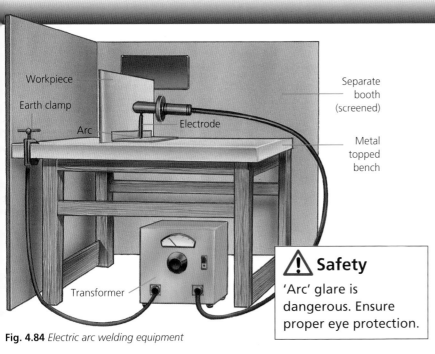

Fig. 4.84 *Electric arc welding equipment*

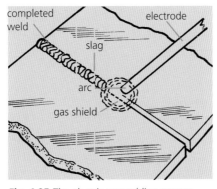

Fig. 4.85 *The electric arc welding process*

> ⚠️ **Safety**
>
> 'Arc' glare is dangerous. Ensure proper eye protection.

In **electric arc welding** (Figure 4.84) an electric arc, of low voltage but high current of 10–120 amps, is struck between a metal electrode and the material to be joined. The **electrode**, as well as carrying current, is a **flux-coated filler rod**. Very intense heat is produced at the end of the arc, melting the electrode and the metals to be joined to form the weld bead. Protection from oxidation is given by the special flux. This generates a gaseous shield, forming a molten blanket over the weld pool. As it solidifies a brittle **glassy slag** is formed, which can be easily chipped away when cold.

Different metal thicknesses require different diameters of electrode and different currents. This process is widely used because of its low capital and running costs.

To prepare the joints, paint, rust and any galvanised (zinc) coating must be removed. Thicker metal requires edge treatment such as bevelling, so that the weld achieves strength by penetration into the metal. **Spot welding** (Figure 4.86) is used commercially to give intermittent welds and some pre-tacking may be necessary with long runs.

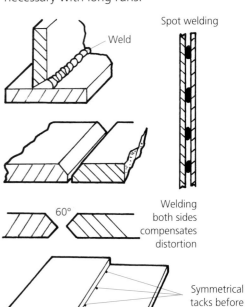

Fig. 4.86 *Jointing*

Aluminium is an important material but is difficult to work because of its oxide film. MIG (metal inert gas) and TIG (tungsten inert gas) are processes used for welding aluminium.

Putting it into practice

Conventional construction techniques are associated with joining together like (similar) materials (e.g. wood to wood, metal to metal). Today's problems increasingly concern the use of **multi-materials**. To achieve the best results it is often necessary to blend traditional construction with contemporary techniques. Therefore when designing a construction to suit a real situation a number of things must be borne in mind.

■ The **function** of your fabrication must be ascertained. Is its purpose to support or to contain, and is a mechanism involved? Remember, it must support the weight and resist a number of different forces acting upon it.

■ The way the fabrication looks is important. Often the solutions themselves influence **aesthetics**. Will the construction be hidden or visible? It must be appropriate to its surroundings.

■ The **materials** used should be carefully selected. Different types offer contrasting properties and methods of working, as does the amount or thickness of available material. Whatever materials you choose you must try to achieve **maximum strength**, using the **minimum material**.

Remember, while it is safe to over-engineer, it is not cost-effective!

■ It is also important to consider the **simplest method of working**, in order to reduce work and save time. Often this means reducing and standardising the number of components. An exception is shown in Figure 4.87, where a solution to the problem of joining multi-strand cable to a metal rod introduces a third party, in order to make it **knock-down**.

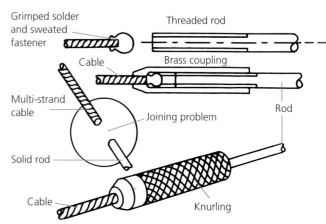

Fig. 4.87 *Joining multi-strand cable to metal rod for a lawn mower cable*

EXERCISES

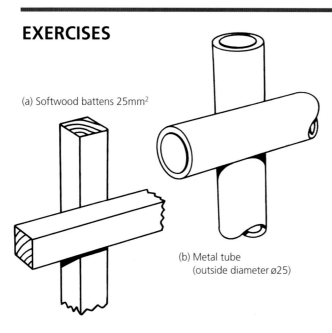

(a) Softwood battens 25mm²

(b) Metal tube (outside diameter ø25)

Fig. 4.88 *Clamping wood battens and metal tube for a temporary garden structure*

1 A temporary framework is to be erected in the garden for growing annual climbing plants such as sweet peas or runner beans. Figure 4.88 shows horizontal and vertical **square wood battens** (a) and **metal tubing** (b) used in the framework. A simple reusable clamp is required to keep (a) and (b) in position, adding to the rigidity of the structure. Sketch four alternative solutions and develop one fully making simple mock-ups, using paper or card, etc.

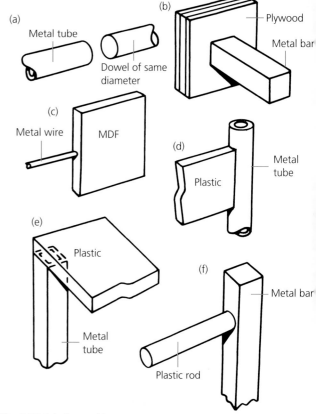

Fig. 4.89 *Jointing problems*

2 In the six jointing problems, (a) to (f) illustrated in Figure 4.89, sketch two alternative ideas for each which provide both a **permanent** and a **knock-down** solution to joining.

5·Materials in action

This chapter contains real examples of project work from schools and colleges. It shows a range of work using different materials and techniques based on a variety of themes and topics.

Making can be the most enjoyable and most rewarding part of Design & Technology project work – it is also the most crucial, and usually the most time consuming. It is helpful to set your work within the context of the whole of designing and making, the two elements that make up GCSE Design & Technology courses. Many people have tried to describe the complex relationship between 'thinking' and 'doing' – between generating ideas and realising them. This has led to many models being created to explain what has become known as the 'design process'.

The design process in its simplest form can be seen as a series of steps, like those shown below. It is extremely important, however, to respond to Design & Technology challenges in a manner that satisfies the requirements of the particular examination or course that you are following. This may vary from the model shown here, but the two key elements, 'designing' and 'making', will be very evident.

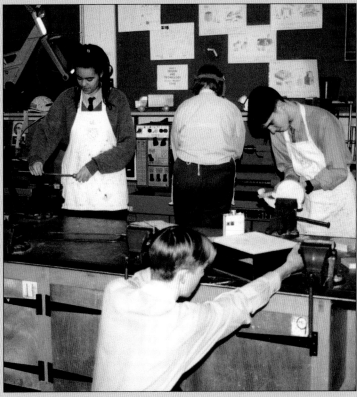

A busy D&T workshop

A 'DESIGN PROCESS' MODEL

Lighting

Toys

Seating

The starting point:
Investigation of a situation, a design brief, or an evaluation of a product or process from which a design specification is developed that describes what needs to be done.

Generating and developing ideas:
Research and analysis leading to annotated sketches and models that help to clarify and visualise what is to be made and show the details necessary to support the planning for making.

Planning:
Consideration of the steps to be taken in realising the design including the materials and components needed and the time to be taken.

Making:
Materials in action

Testing and evaluation:
Testing and evaluation against the design specification and the original intention. Can improvements be made, and what lessons have been learned and experiences gained?

Jewellery

Concept design

Large structures

Focus on lighting

Lighting can provide many opportunities for interesting and creative project work – it lends itself to a wide range of design ideas, materials and processes.

Lights can be specifically designed for a particular function or situation, or fun lights can be designed for particular people or age groups. It is always important to explore the way in which the light interacts with the material used; the opacity (opaqueness) of the material is critical. However, remember that lights usually spend most of their time switched off, so they also need to look good and have a strong aesthetic appeal when they are unlit. Colour, geometry and symmetry are important considerations, along with stability and electrical safety.

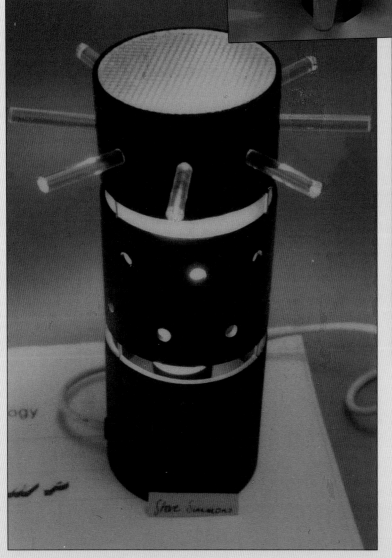

Many of the examples on this page and the next make use of acrylic, which is available in a wide range of colours as well as clear, tinted and opaque.

The three acrylic lamps shown above have been designed to appeal to children. The lamp on the left makes use of the effect of light shining through holes and the 'fibre optic' effect of light passing along clear plastic rods.

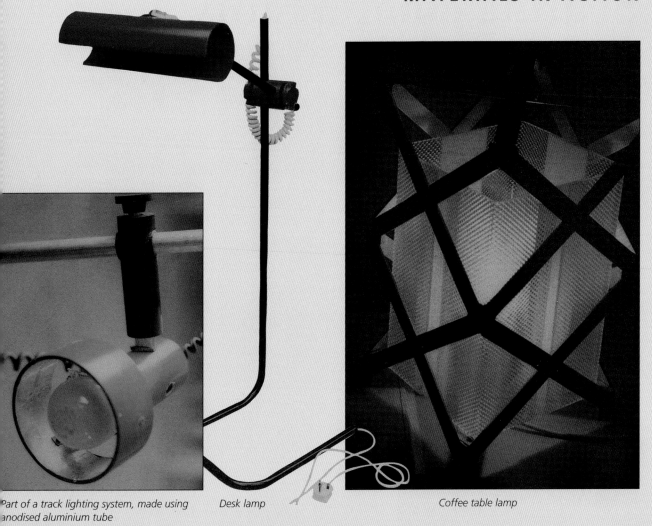

Part of a track lighting system, made using anodised aluminium tube

Desk lamp

Coffee table lamp

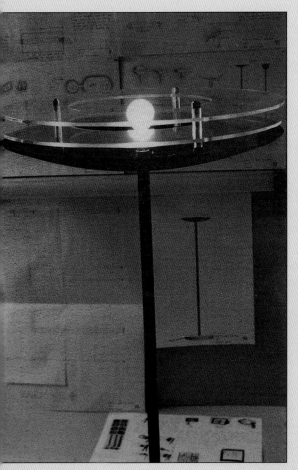

Standard lamp

This interesting lamp (above) has been made using a computer controlled milling machine to cut out several identical acrylic shapes

Seating comfortably

Whatever their function, whether for work or relaxation, seats need to be designed ergonomically. They must always be comfortable yet also help the user in whatever they are doing. A seat that is intended for use when working at a keyboard has to be different from one intended for use when watching TV. Ergonomics deals with the way that people interact with the things that they use and work with. In the case of a seat, consideration has to be given to many human aspects – the length of the user's upper and lower legs, the support required for the lumber region of the back, and the position of the support for the head and arms if these are to be included.

In order to ensure that a seat functions properly, it is a good idea to build a full size model – this also enables you to try out seat padding and covering. The photographs on this page show one particular student's project being prototyped in this way.

The photographs on the opposite page show an interesting range of seats designed and made by students at one particular school. Seats often use a combination of materials, providing many challenges for fabrication. The materials seen here include: wood, manufactured board, metal and fabrics. It is also worth considering how these materials work together aesthetically – too many materials in one project seldom harmonise or look good together. Seating in the home and in public places has provided many famous designers with opportunities to extend their work; it is worth bearing this in mind when researching seating projects.

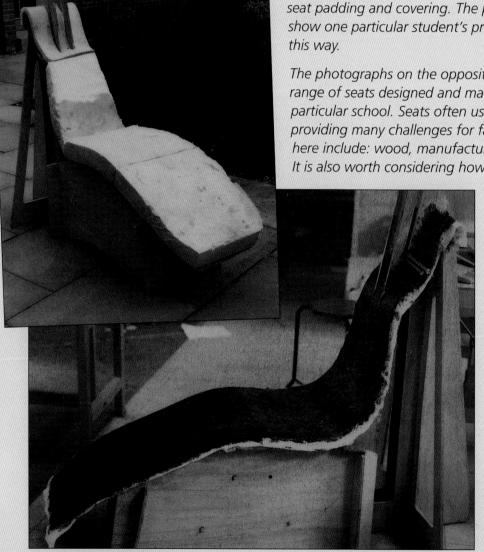

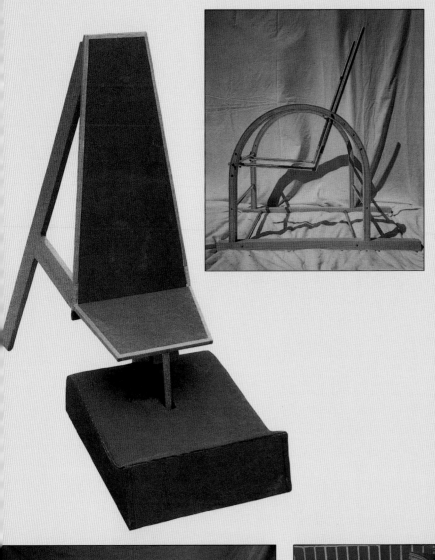

Concept design

Concept design is about developing new ideas and new conceptual forms through models and prototypes. It is about turning unusual ideas into reality or into a form that enables them to be explored further through evaluation and testing. Concept design provides the opportunity for diverse thinking, for stimulation and for fun.

Concept car design is used by car manufacturers to try out ideas. Because the ideas are so new, they usually influence the design of sports cars first, and later go on to affect family car design.

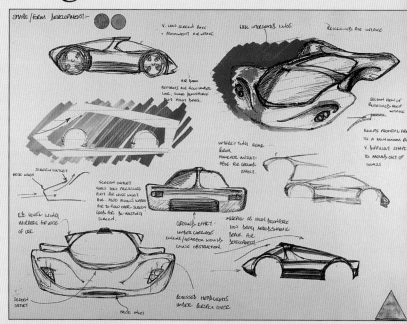

Design drawings

This concept car was realised using GRP in a plaster of Paris mould formed from a clay, polystyrene and wood pattern. In the photographs you can see the pattern being built up, and in its finished clay form. The plaster of Paris mould was built around the pattern.

Polystyrene foam and wood provides the skeletal form for the pattern

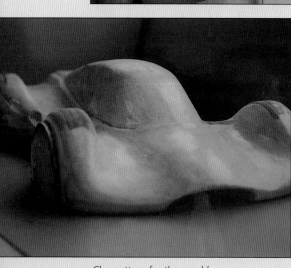

Clay pattern for the mould

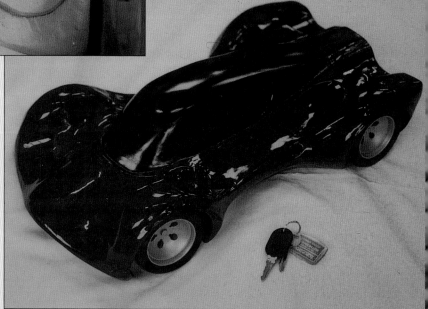

The finished concept design

Cycling enthusiasts have experimented with the concept of fun cycles and recumbent (lying-down) cycles for many years.

In the project shown on this page, the student has taken this approach to pedal powered transport and endeavoured to turn it into reality. The material used is tubular aluminium, which provides both lightness and strength, but is a difficult material to weld.

'Lego' model to explore the chain drive and the steering

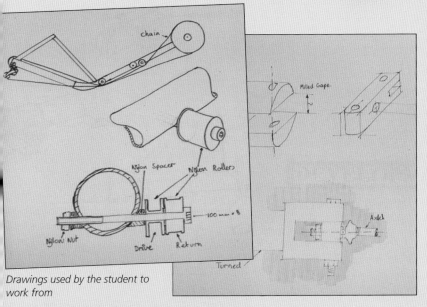

Drawings used by the student to work from

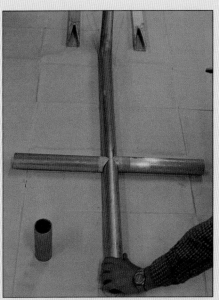

Detail of the frame during manufacture

The finished product

Toys

Projects based around toys and games provide opportunities to work with a wide variety of ideas and materials. Toys can incorporate mechanical and electronic systems, and the outcome can be as imaginative as you are. This does not mean that it is necessarily an easy area to design for – there is much to consider. What is the target age group? Does the toy have an educational value? Will it sustain the interest of the user? Is it safe to use, with no small parts that can come off and be swallowed etc?

Toy manufacturers have a very strict code of conduct with which they must comply. They also have to balance the traditional approach to toy making, using wood and simple shapes, with the ability to capitalise on the latest trend that is popular on the television or cinema.

Consider the qualities that a toy should have and evaluate the strengths and weaknesses of toys that you know about and the examples shown here.

The two 'traditional toys', above and on the right, have been made from brightly painted wood. The 'push along' duck has a cranked axle with a stiff wire connection to its beak to make it appear to 'quack' as it goes along.

The cam action that operates this 'bobbing head' toy can be seen clearly. Cams are a simple method of building some motion into toys; either by turning a handle, as in this case, or by using the rotation of an axle.

The car and the yellow submarine in the photos below are examples of toys that have electronics incorporated into them. These two toys are both money boxes with a difference. When coins are dropped into them an electronic circuit is completed which makes them flash lights and make a noise. The car is constructed from GRP and the submarine is sheet acrylic.

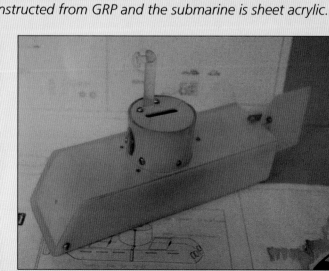

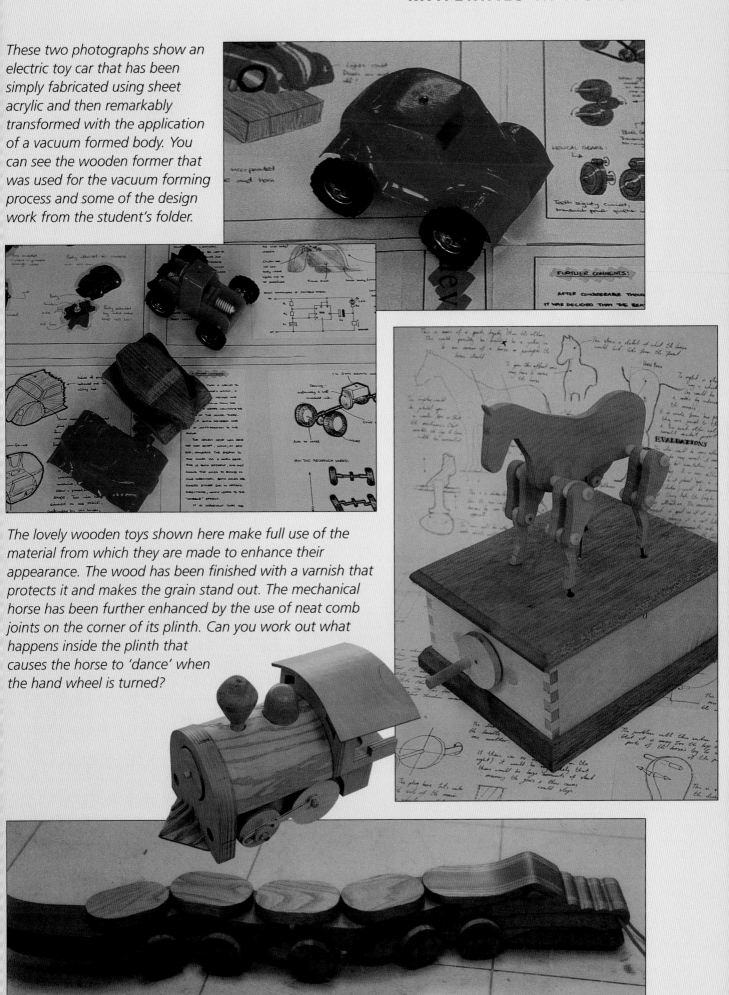

These two photographs show an electric toy car that has been simply fabricated using sheet acrylic and then remarkably transformed with the application of a vacuum formed body. You can see the wooden former that was used for the vacuum forming process and some of the design work from the student's folder.

The lovely wooden toys shown here make full use of the material from which they are made to enhance their appearance. The wood has been finished with a varnish that protects it and makes the grain stand out. The mechanical horse has been further enhanced by the use of neat comb joints on the corner of its plinth. Can you work out what happens inside the plinth that causes the horse to 'dance' when the hand wheel is turned?

Jewellery

Jewellery making provides an opportunity to explore the full potential of materials – how they 'work', their appearance and how they can interact with other materials such as glass and stone. Designs can come from a wide variety of starting points – from natural forms and from social, economic and historical perspectives. The student's work represented here is a good example of this. She began by studying natural forms, in particular seed pods, and by linking this to a study of Scandinavian, particularly Viking, culture. Some of the examples have, in fact, led to work being commissioned by a Norwegian jewellery manufacturing company.

On the right you can see some development ideas for the forged silver ring shown in the photograph.

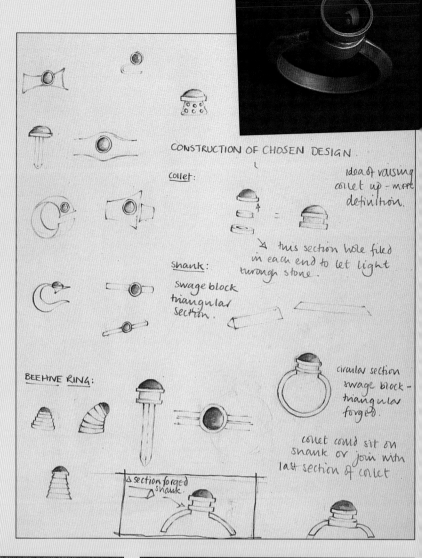

CONSTRUCTION OF CHOSEN DESIGN.

collet:

idea of raising collet up - more definition.

this section hole filed in each end to let light through stone.

shank:
swage block triangular section.

BEEHIVE RING:

circular section swage block - triangular forged.

collet could sit on shank or join with last section of collet

△ section forged shank.

POD FORMS:

Twisting forged silver around pod to support it, hold contents → peas!

The 'Pod' pendant, shown here with some of its design development ideas, was inspired by a study of plants, like peas, that contain their seeds in pods. The pendant is made from copper and opens up to reveal a bracelet contained within it.

Designing and making jewellery in metal allows many processes to be used. Large-scale processes such as casting can be miniaturised, and processes like forging simplified. Careful and precise hand-working using fine saws and files is crucial. It is also very important to concentrate upon finishing and polishing to achieve the desired quality of finish.

A wide range of finishing processes can be applied, many of which have been developed by jewellery makers over the centuries. The examples of work seen on these three pages include flypressing, etching, stone-setting and planishing, employed as finishing techniques for decoration and protection. Planishing also hardens the metal by stiffening and strengthening it.

The necklace featured on this page is made from cast pewter on a wire form. Each 'tooth' had to be carefully hand finished and polished.

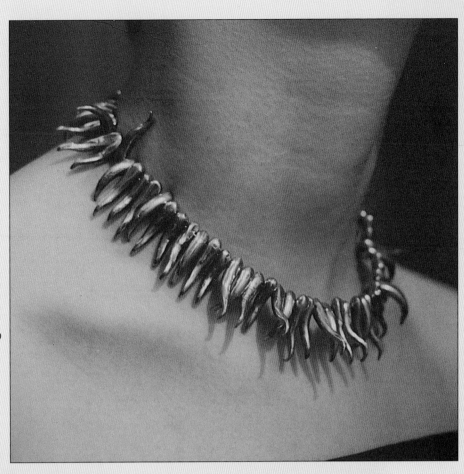

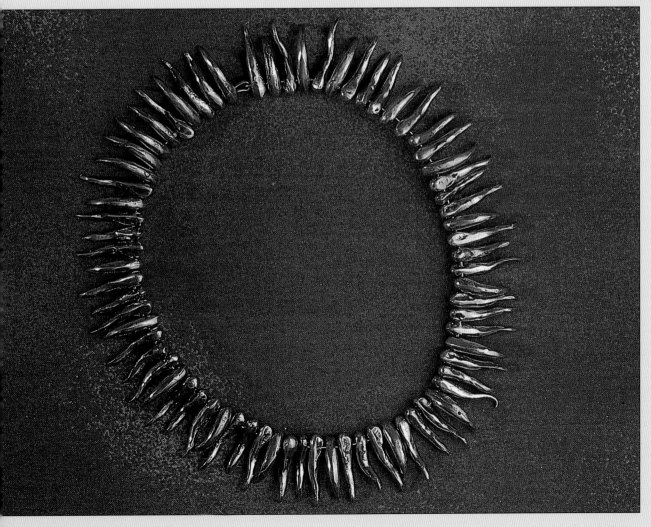

The work shown on this page has developed into the commercially produced range of necklaces, as seen below.

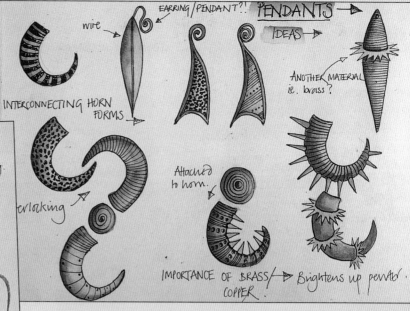

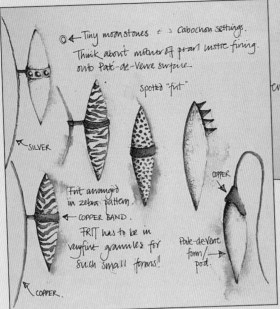

The material used is pewter and the manufacturing process is centrifugal casting. This process involves pouring molten pewter into a rotating silicon rubber mould that is shaped like a wheel. The molten metal is poured into the centre and centrifugal force pushes it into the mould cavities that are near to the rim.

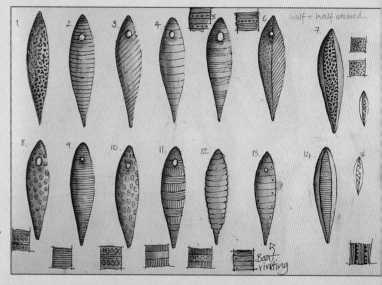

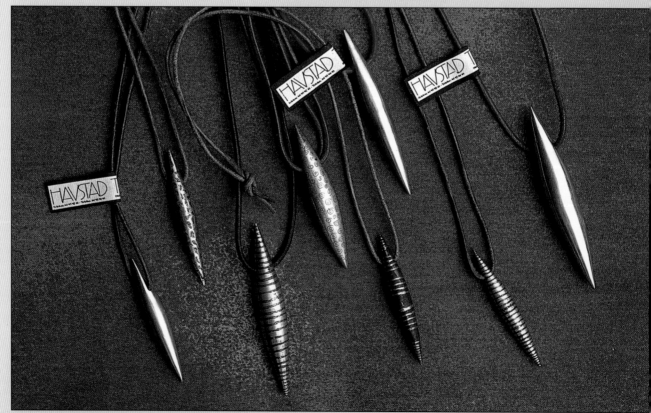

Large structures

Large structures present particular problems, especially in terms of space, time and cost. Schools rarely have enough storage space so large projects manufactured within school workshops should be either designed to be 'knock down' or they need to be carefully planned so that they only come together fully at the end of the project. It is critical, therefore, to consider carefully the fabrication techniques that are going to be used. You also need to develop the ability to 'sub-problem' – this means breaking down a large problem into its component problems and tackling them in a systematic and sequential manner.

Time is always a problem for projects of any kind, but with large projects it is very easy for small problems to become major hold ups. You need to establish very early on where, when and how you will need to access the material that you require. Large quantities of tubular materials, for example, may have to be ordered specially, which takes time and can cause delays. Also, you need to carefully consider size and weight – there could be some major embarrassments if the outcome of hours of work is too big to go through the door or too heavy to move!

The trailer in the photos on this page is of a modular construction made from tubular aluminium and steel (at the top of the page you can see some of the construction in detail). It can be used for carrying bicycles and as a flat-bed or general purpose trailer. Most of the structure has been bolted together, but there has also been some welding. The natural appearance of the wood 'planking' on the sides has been enhanced using a decorative preservative.

This large structural project is a wind powered generator. A project of this nature presents a number of problems, and it was helpful to consider them as sub-projects that contribute to the whole: (a) capturing the wind via the blades (wind vanes), (b) generating electricity, and (c) making a stable, strong and portable structure that is in fact constructed in three sections.

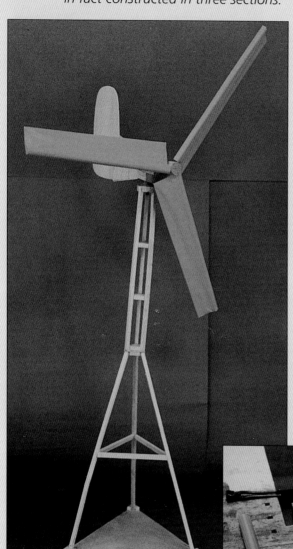

The project was first modelled using balsa wood to test the triangulation and modular construction of the tower. You can see the model in the photo on the left. In the photographs below you can see the covering of the blades being 'ironed' to make it shrink tight on to the form of the blades.

On this page the photographs show the construction of an alternative method of utilising wind power. This type of device is known as a Savonious rotar.

It is 1.5 metres tall and is made from hardwood, plywood, steel rod and carefully formed sheet aluminium. The photographs show the stages in construction, using a wide range of fabricating techniques and manufacturing processes.

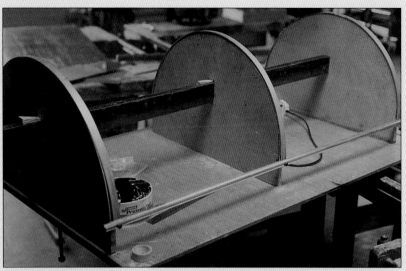

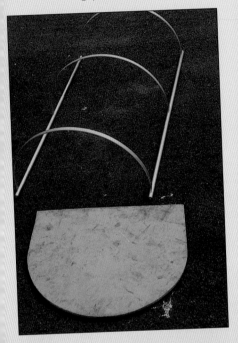

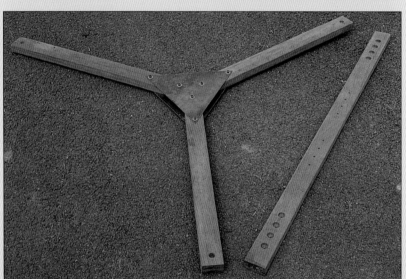

Acknowledgements

The publishers would like to thank the following:

For their help in providing case study material:

The Health and Safety Executive; Roy Walker and Bob Preece; Richard Pullen; David Jones, Steve Currier, John Horton and Fred Sharpe at GKN plc; Brian Illingworth at RPC Containers Ltd; Ray Picker at John H. Rundle Ltd; Roger Newborough at Jolly Roger (Amusement Rides) Ltd; Dave Thomas at Auto-Trail Ltd and Vanroyce Caravans; Laura Baxter; students and staff at De Aston School.

For their help with location photography:

William Farr C of E Comprehensive School, Welton (Head of Technology: Mike Finney); De Aston School, Market Rasen

For permission to reproduce photographs and illustrations:

Alton Towers (2.5)
Armitage Shanks Ltd (2.78, 3.113)
Autodesk Ltd (1.15, 1.16, 1.17, 1.18)
Bates & Lambourne (2.24)
Black & Decker (3.31)
BMW (2.4, 4.6 right)
Boxford Ltd (1.13 left, 3.23, 3.130)
British Aerospace (1.4)
British Motor Industry Heritage Trust (1.49)
British Steel (1.60)
British Telecommunications plc (2.79 right, 2.80)
Canary Wharf Ltd (2.2)
Colin Chapman (2.72a, 3.5, 3.85, 3.86)
Christie's Images (3.160)
Chubb Safe Equipment Company (1.59)
C.R. Clarke & Company (UK) Ltd (3.94)
The Consortium (3.83)
Cybernetic Applications (1.42, 1.43)
DCA Design International Ltd (1.13)
Design and Print/Tim Guthrie © (2.50)
Draper Tools Ltd (1.19 [ii], 3.15, 3.22, 3.26, 3.29, 3.37 middle and bottom, 3.53, 3.54, 3.55, 3.56, 3.62, 3.140, 4.32, 4.33)
Dupont (1.11)
Elm Energy & Recycling (UK) Ltd (1.70)
Environmental Picture Library (2.26)
European Gas Turbines (1.58, 2.18)
F.W. Chandler Ltd (2.42)
GKN (1.19 [iii], 1.77)
Goodfellow Cambridge Ltd (2.92)
Graduate Lathe Company Ltd (3.2)
Habitat UK (2.73)
Hammersmith Hospital (1.3)
Hepworth Building Products (3.116)
Her Majesty's Inspectorate of Pollution (1.71)
Hulton Deutsch (1.1, 1.50)
Hygena/MFI (1.12)
IBM UK Labs, Hursley © (2.3)

ICI (1.4)
The Image Bank (4.5, 4.14, 4.17)
Jolly Roger (Amusement Rides) Ltd (3.1)
Kenwood (1.10 left)
Machine Tools Technologies Association (1.36)
Magnet Ltd (3.81)
Massey Ferguson (1.35, 1.47, 1.48)
The Merry Hill Centre (1.75)
Moore & Wright (3.26)
National Dairy Council (3.118)
National Trust Photographic Library (2.23, 2.25)
Nissan (1.5)
Mel Peace (2.74, 2.89, 2.90, 2.91, 3.69 photo, 3.77, 3.91, 3.93, 3.103, 3.144, 3.154, 3.156, 3.161, 3.163, 3.164, 3.166)
Peugeot Partnership Centre (1.65)
Fredk. Pollard & Co Ltd (1.33 right, 1.37, 1.38)
Quadrant Picture Library (1.10 right, 2.13)
Record Tools (3.27, 3.28, 3.32, 3.84, 3.143)
Roland Digital Group (1.14)
Royal Botanic Gardens Kew (2.1)
RPC Containers (2.85, 3.114)
Science Photo Library (1.2, 1.57, 1.66, 2.10, 2.11, 2.22, 2.30, 2.36, 2.59, 2.61, 4.4, 4.9, 4.11, 4.16, 4.26)
Peter Sharp (1.19 [i], 3.88, 3.159)
Sony (1.10 top)
Stanley Tools (3.42, 3.46)
Stäubli Unimation (1.39, 1.44)
Tebrax Ltd (4.61 photo)
Telegraph Colour Library (1.68, 3.4, 4.1, 4.10)
Tony Stone (1.19 [iv], 2.70, 4.7, 4.8, 4.20)
Unilever (1.4)
V&A Picture Library (2.79 left)
Jacolyn Wakeford/ICCE (2.38)
Alison Walters (2.87, 3.37 top, 3.117, 4.18)
Warren Machine Tools Ltd (3.131)
John Wilman Ltd (3.115)
Zanussi (4.6 left)

Index

Index

Index